D1143227

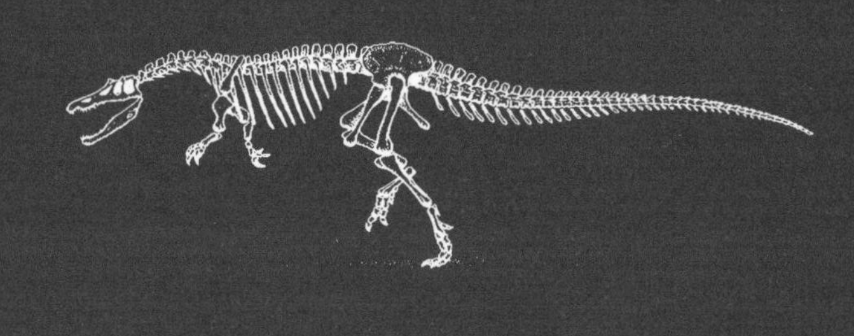

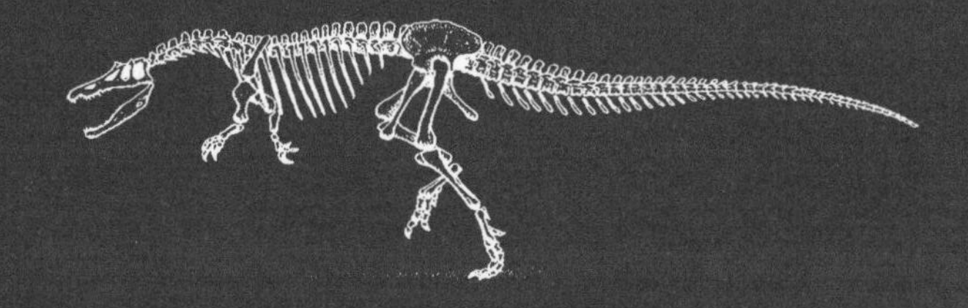

THE NATURAL HISTORY MUSEUM BOOK OF DINOSAURS

THE NATURAL HISTORY MUSEUM BOOK OF DINOSAURS

TIM GARDOM & ANGELA MILNER

CARLTON
BOOKS

THIS IS A CARLTON BOOK

This edition published by Carlton Books Limited, 2006
20 Mortimer Street
London
W1T 3JW

First published in 1993

A CIP catalogue record for this book is available from the British Library

ISBN-10: 1-84442-183-X
ISBN-13: 978-1-84442-183-1

Printed and bound in Dubai

CONTENTS

INTRODUCTION

Dinosaurs are a source of enormous fascination and ever-growing popularity with people of every age from three to adult. Sometimes it seems as if, 65 million years after they disappeared, the world is once again being taken over by them.

Why are they so fascinating? Probably because they seem so amazingly different from the animals we know – bus-sized plant-eaters, hunters with 20 centimetre-long serrated teeth, strange creatures with outlandish names. Yet if we only think of dinosaurs as monsters we will never get past the 'Did you know? Amazing Facts!' level of knowledge about them. The really compelling fascination of dinosaurs is not how different but how similar they are to the animals we know.

Despite their monstrous reputation many dinosaurs were, in fact, no larger than animals that can be found living today. All of them had to meet the same basic challenges of survival. In their search for food and space, their methods of moving and eating and their social organization, dinosaurs show striking similarities to modern groups, particularly birds and mammals. Over the last fifteen years spectacular new discoveries made in the field and laboratory have demonstrated even more clearly how much dinosaurs and today's animals have in common.

This book is a natural history of dinosaurs. It has grown out of the permanent exhibition at The Natural History Museum in London. The exhibition and this book look at dinosaurs as living creatures in a real environment. Drawing on evidence from around the world, the dinosaurs are shown to be a varied, balanced, dynamic community and one of the most spectacular success stories in the history of life on Earth. This enlarged and updated edition includes up-to-the-minute results of cutting edge research, detective work that helps us to understand even more about dinosaurs and their living relatives that are all around us.

THE DINOSA

I

Dinosaurs are one of the great success stories of evolution, utterly dominating the land for 160 million years. Yet though they inhabited a world very different

JR WORLD

from our own, they had much in common with today's animals. Living in balanced communities, they faced the same survival needs — to breed, to feed, to hunt and escape their enemies.

The Earth is constantly changing. Its climate, the shape of seas and continents, even the position of the magnetic pole have all changed many times during its 4600 million year history. These changes have determined the evolution of life, creating an environment that favours some plants and animals more than others. For a while the favoured life forms thrive and diversify, while other, less adaptable forms, die out. Then the environment changes again and new forms of life take over the planet.

It used to be thought that these changes weeded out the weaker life forms and allowed the better ones to press ahead in a constantly improving march of progress. The phrase 'survival of the fittest' best sums up this idea. In fact, the surviving forms of life are only 'better' in the sense that they can cope with the environmental changes the Earth throws at them. At its simplest, if the climate turns cold, animals with fur stay warm and do better than those without any insulation. Yet fur is not 'better' than scaly skin, just useful in different conditions.

Dinosaurs have suffered particularly badly from the 'survival of the fittest' prejudice. The fact that they became extinct was seen as proof that they were not 'fit' to survive. They were too big, too slow and too stupid. This is wrong on two counts. Firstly, not all dinosaurs were big, slow and stupid. There were plenty of small, fast, clever ones as well. More significantly, although dinosaurs eventually did become extinct, they survived for a huge span of time — 165 million years — more than forty times longer than Man has so far achieved. Indeed, to say that they survived does not do justice to their evolutionary achievement. The fossil evidence clearly shows that, while they were alive, every land habitat worth having was occupied by some kind of dinosaur. Flying reptiles, and later birds, were able to colonize the air where the dinosaurs could not reach them. Huge marine reptiles swam in the dinosaur-free seas, but on land, with very few exceptions, no animal larger than a domestic cat ever managed to evolve while the dinosaurs were around. The early mammals, which are our ancient ancestors and whose superior intellectual powers we hear so much about, were mostly rat-sized creatures that scuttled about at night when the dinosaurs were safely asleep. Dinosaurs utterly dominated the land in a way that no other group of animals had done before.

In this book we set out to celebrate the dinosaurs' success and, as far as possible, to explain it. This means looking at dinosaurs not simply as giant monsters or complex skeletons, or even as fascinating twigs on a family tree, but as real live animals in a real environment, as members of a balanced animal community facing the same survival needs as today's animals — to breed, to grow, to eat and avoid being eaten.

THE EVIDENCE

We have two main sources of information. The first is the fossil record — the physical evidence of the dinosaurs' existence left behind in rocks

HOW ARE FOSSILS FORMED? Dinosaur dies and is rapidly covered with water-borne sediment. Sediment compresses and hardens. Minerals seep through the bones which gradually change into rock. Millions of years later the fossil appears as the covering rock is worn away by wind and water.

(Previous spread) SURVIVAL OF THE FITTEST. Three *Deinonychus* attack a solitary plant-eating *Tenontosaurus* that has unwisely strayed from its herd.

around the world. The word fossil comes from the Latin *fossilis*, meaning 'dug up', and originally meant anything that was taken out of the ground, including minerals and crystals. Today, a fossil means the traces of any past life preserved in the rocks.

Despite surviving for so long, dinosaurs represent only a small fraction of the fossils found on Earth. Usually when an animal dies its body is rapidly destroyed. Scavengers tear pieces off and carry them away, the flesh rots, the skeleton falls apart, the bones disintegrate and within a few weeks there is no recognizable trace. If the body is quickly covered with some sort of sediment, however, this process of disintegration may not occur. A desert sandstorm could do this, or mud on the banks of a river. If the dead animal fell into deep water or was washed into a lake, then the silt at the bottom could quite quickly cover it over. This is why by far the most numerous fossils are those of sea animals, particularly those that lived in shallow water near the coast, where mud and silt were constantly present to bury their dead remains. It is also why the bodies of dinosaurs (all land animals) very seldom became fossilized. It took unusual circumstances for the bones of a dinosaur to avoid the common fate of scattering and destruction. However, some dinosaur remains were covered rapidly and then turned into fossils in a number of different ways.

Once it had died and the sediment had covered it, the flesh and other soft parts of the dinosaur almost always rotted away, leaving only hard bones and teeth behind. Gradually, the sediment built up on top of the bones, compacting into rock such as limestone, mudstone, sandstone, clay or shale. As this happened, minerals from water percolating through the surrounding rocks seeped into the bone structure, impregnating the tiny spaces and often altering the original mineral in the bone. These fossils, part original bone and part rock, are called 'petrified'.

Sometimes acidic water in the ground dissolved the bone and left a hollow mould where it used to be. These are 'natural mould' fossils and, by pouring material such as plaster of Paris or rubber latex into them, the precise shape of the bone that used to be there can be created. In other cases, these natural moulds filled later with sediments or quite different minerals such as silica, calcite or iron pyrites which gradually built up a perfect replica, creating a 'natural cast' fossil. The rarest fossil of all formed when the dinosaur's body had been covered in a dry environment and some of the soft parts had become mummified and then fossilized.

In these fossils the texture of the skin, and even the folds in it, can be clearly seen. The colour is not preserved, however, since the surrounding rocks give their own colour to the fossil.

In addition, there is a wide range of other fossils, not actual dinosaur remains but evidence of their presence and their way of life (what police would call 'scene of crime' evidence). These include fossilized footprints or trackways, nests and eggs, scratches in the ground, tooth marks on bones, fossilized dung, stomach stones and much more besides. Taken

OTHER EVIDENCE. Apart from fossil bones and teeth, dinosaurs left other clues about their lifestyle in the rocks, such as this fossilized *Oviraptor* egg from the Gobi Desert.

TRACKS AND TRAILS. Footprints and trackways tell us whether dinosaurs lived alone or in groups, and how fast they moved.

altogether these direct and indirect remains make up the fossil record — the hard evidence (literally) on which our knowledge of dinosaurs as living animals is built.

The other major source of information about dinosaurs is observation of animals alive today. It is no accident that the first dinosaur scientists were also the world's leading experts on comparative anatomy, the study of animal structure to determine relationships between species. Their knowledge of animal anatomy allowed them to see the similarity between dinosaurs and some living creatures such as birds, lizards and mammals. They could also appreciate the difference of scale that made the dinosaurs so extraordinary. So today's animals are an essential point of reference. This not only covers basic issues such as how dinosaurs stood and walked, but also questions about their behaviour, their communication, their colour and their techniques for hunting and defence. Throughout this book we will be comparing what we know about dinosaurs with observations from our modern world.

Where did dinosaurs come from? That apparently simple question has been the subject of intense debate amongst scientists for over 150 years, and even today there are widely differing views about the origin of dinosaurs.

Life had been present on Earth for at least 3260 million years before the dinosaurs put in an appearance. Starting with simple one-celled organisms, over hundreds of millions of years the sea came to swarm with a huge variety of animals including worms, jelly fish, shelled molluscs and, later, fish with bony skeletons. The land was also gradually colonized, first by simple single-celled plants such as algae, and later by more complex plants and by animals — worms, arthropods (insects, scorpions, mites and others) and molluscs. One group of bony fishes, the lobe-fins, which had lungs and strong fins with bony internal supports, began to venture on to land about 370 million years ago. These pioneers were the ancestors of both land-dwelling amphibians and the early reptiles.

By 245–235 million years ago there were many groups of reptiles living on the land. These included mammal-like reptiles that gave rise to mammals, rhynchosaurs (plant-eaters with curved beaks and strong digging hind feet), and archosaurs (the 'ruling reptiles'). The earliest archosaurs, informally known as thecodontians or 'socket-toothed animals', were all meat-eaters. Some were large crocodile-like animals with sprawling legs. Others developed a semi-sprawling stance and a special rotary ankle joint. This meant that they usually moved with their legs bent at right-angles at the elbow and knee, but were able to run over short distances with

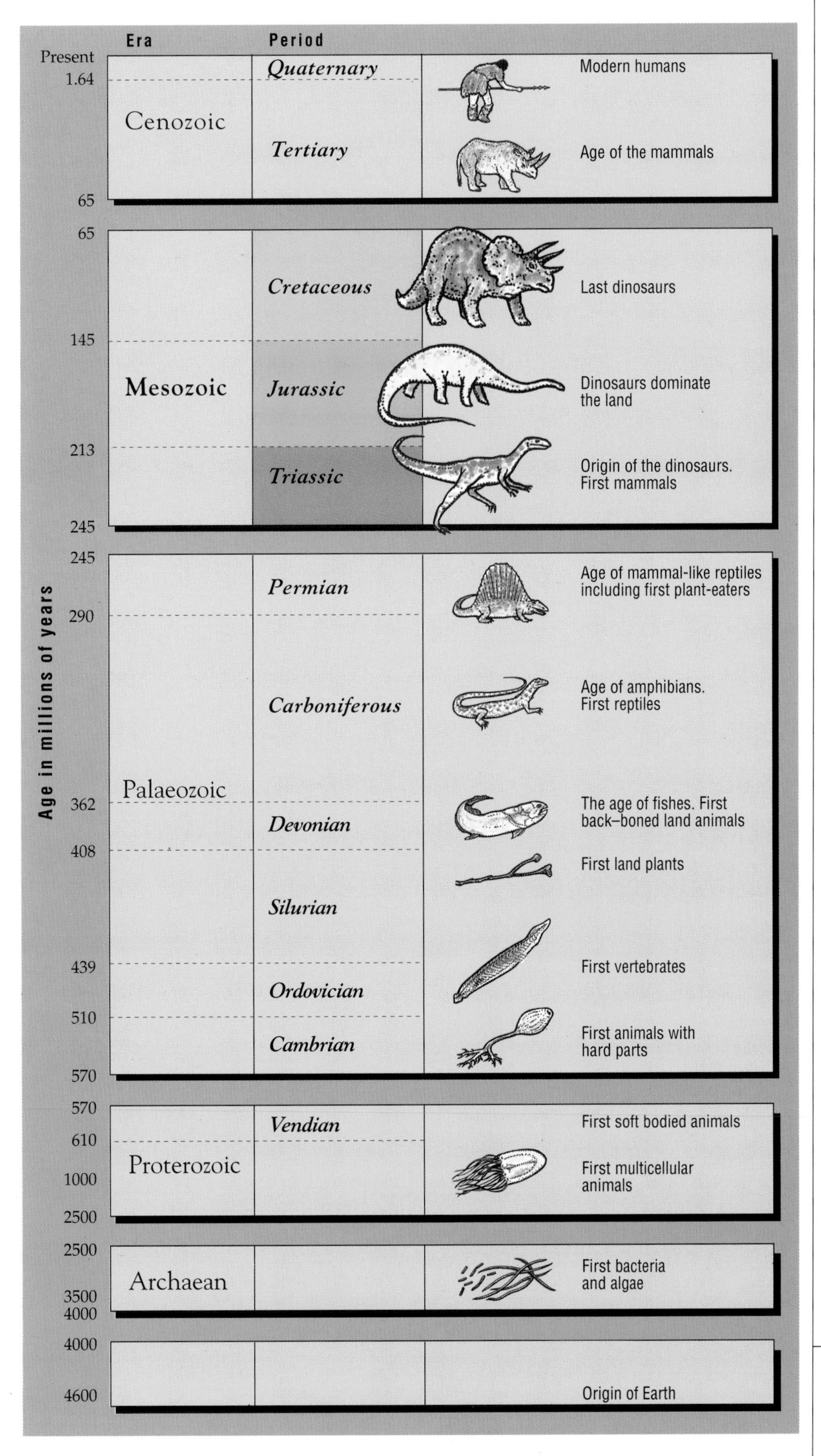

their legs almost straight (modern crocodiles are their descendants). Smaller, lightly-built thecodontians were the first animals to develop the ability to run for short distances on their back legs and for some this became the fully improved stance, with straight legs permanently tucked under their bodies. The little 30 centimetre-long thecodontian *Lagosuchus* from Argentina stood on the anatomical boundary between these improved stance thecodontians and the two groups that evolved from them — the flying reptiles (pterosaurs) and the dinosaurs themselves.

The dinosaur era, officially known as the Mesozoic Era, is divided into three parts called the Triassic, Jurassic and Cretaceous Periods. These divisions do not really have anything to do with the evolution of the dinosaurs themselves, which carried on developing undisturbed between one period and the next, but relate to the different ages of rock where their fossils are found. The three periods are, however, essential reference points for dinosaur scientists, who usually talk about the early, middle or late part of each to define when a particular dinosaur lived. During the Mesozoic Era the Earth and its environment underwent changes that had a profound effect on how dinosaurs evolved.

THE TRIASSIC PERIOD

In the Triassic Period (245–213 million years ago), all the continents on Earth were joined in one giant land mass. Called Pangaea (meaning 'All Earth'),

IN THE BEGINNING... Dinosaurs evolved over millions of years from earlier life forms. Their nearest living relatives today are birds and crocodiles.

STEPS ON THE WAY. Early archosaurs with sprawling gait (1) gave rise to semi-improved walkers with rotary ankles (2). Small, lightly-built archosaurs like *Lagosuchus* (3) developed a fully upright stance and became bipedal. They gave rise to the earliest dinosaurs such as *Herrerasaurus* (4) and *Eoraptor* (5), the most primitive meat-eating dinosaur known. All the theropods evolved from similar-looking ancestors.

this supercontinent straddled the equator. The temperature on land stayed constantly warm and, with no polar ice-caps or large inland seas to affect the climate, there was very little seasonal variation, although it became hotter and drier towards the end of the Triassic. There seems to have been quite a difference in the plant life between the north and south of Pangaea although scientists are not clear why this should be so.

Because Pangaea was a single continent the dinosaurs were able to spread across it quite freely. Fossil remains of similar kinds are found in different parts of the world. The very earliest dinosaurs were small, slightly built, two-legged meat-eaters. *Eoraptor* is the most basal predatory dinosaur known; it had unusual leaf-shaped teeth at the front of the upper jaw and was only about 1 metre long. Other early dinosaurs, like *Herrerasaurus* and *Coelophysis*, up to 3 metres long, were swift and vicious with sharp, pointed teeth and grasping claws. The first plant-eating dinosaurs also appeared at this time in the shape of prosauropods like *Massospondylus* and *Plateosaurus* which could walk on two or four legs, presumably rearing up to get at higher vegetation. Their fossil remains are very widespread, so their efficient digestive system was obviously able to cope with the differing plant life that spread across Pangaea. Although nowhere near the size of the later giant plant-eaters, the prosauropods were by far the largest land

WHAT IS A DINOSAUR?

Dinosaurs lived between 230–65 million years ago. Dinosaurs lived on land but could not fly. Dinosaurs had straight legs tucked underneath their bodies. Dinosaurs were reptiles. Which of these animals was a dinosaur? (Answers on next page.)

1

2

3

4

5

6

7

8

9

10

THE TRIASSIC LANDSCAPE. The dinosaur world in Europe around 210 million years ago. Giant horsetails grew in wetter areas, tree ferns and bennettitaleans were common, and tall conifer trees grew in drier areas.

THE JURASSIC LANDSCAPE. The dinosaur world in East Africa around 150 million years ago. Extensive fern land and forests of tall conifer trees such as sequoias and monkey puzzles dominated the lush tropical landscape.

THE CRETACEOUS LANDSCAPE. The dinosaur world in western North America around 70 million years ago. Flowering trees and shrubs, similar to beeches, oaks, grapes and hickory, grew alongside cycads, conifers and ferns.

animals of their time (*Plateosaurus* was up to 8 metres long), and an early example of how some dinosaurs used their sheer size to make themselves immune from attack.

The rise of the dinosaurs during the late Triassic led inevitably to the decline of other animal groups that had been very successful previously. Many of the sprawling reptiles and amphibians disappeared as did the advanced mammal-like reptiles. Among the species that cohabited with the dinosaurs were insects of many kinds (often food for small and baby dinosaurs), crocodiles and very small mammals. In freshwater rivers and lakes there was an abundance of life with frogs, turtles and fish, but it seems unlikely that many of these were hunted by dinosaurs. Gliding lizards and the first pterosaurs took to the air and in the sea swam a variety of reptiles like ichthyosaurs and protorosaurs.

The late Triassic can best be seen as a period of colonization by the dinosaurs as they spread over the whole Pangaea supercontinent, replacing the existing inhabitants. Obviously the dinosaurs' superior mobility and speed played an important part in this, although some

A CHANGING WORLD. *Top:* Triassic. All continents joined in the single land mass of Pangaea. *Middle:* Jurassic. Two supercontinents of Laurasia (north) and Gondwana (south). *Bottom:* Cretaceous. Supercontinents continue to move apart and break up, and the land outlines come to resemble those we recognize today.
KEY: yellow = land; light blue = shallow seas covering continental shelves; dark blue = deep oceans.

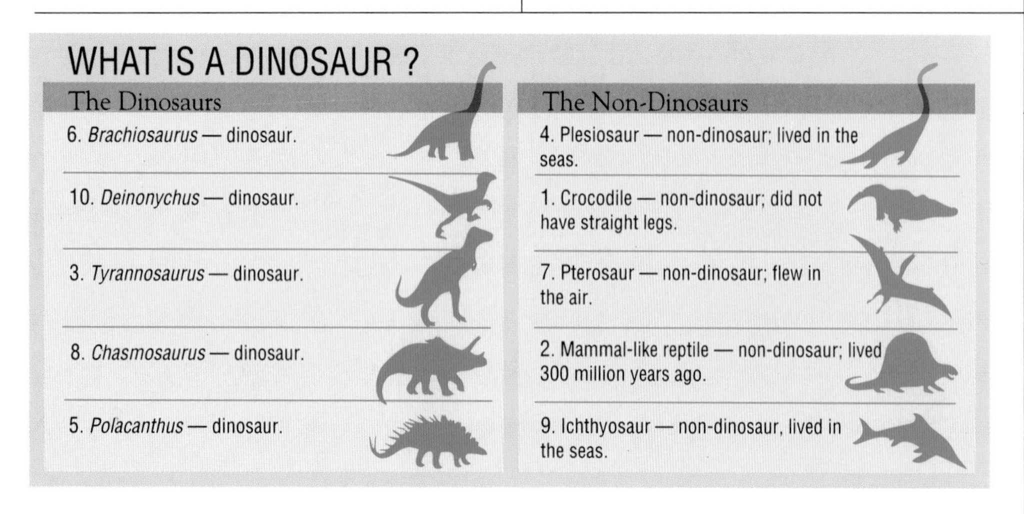

scientists believe that climate changes may have caused many species to become extinct, leaving a gap in the ecosystem which the dinosaurs were able to fill.

THE JURASSIC PERIOD

In the Jurassic Period (213–145 million years ago), the single land mass of Pangaea was split in two by huge rifts in the Earth's surface that created two new land areas, Laurasia in the north and Gondwana in the south. Despite this dramatic separation, fossil skeletons of *Brachiosaurus* and plated stegosaurs are now found as far apart as Africa and North America, showing that there must have been at least occasional land bridges between the new supercontinents.

The break up of the land mass and the creation of large seas between the supercontinents affected the global climate. The temperatures fell slightly and increasing rainfall and a mild climate allowed lush tropical vegetation to grow over huge areas. Ferns and horsetails provided ground cover, while ginkgoes and tree ferns were found near many rivers and lakes. Some of this vegetation became coal seams which we still mine today. Forests of cycads, conifers and sequoias covered thousands of square miles of the drier areas.

The lush conditions gave rise to quite new and extraordinary kinds of dinosaur, among them the sauropods like *Apatosaurus*, *Diplodocus* and *Brachiosaurus*, the largest animals ever to walk on Earth. Anything that needs to consume a tonne of greenery a day has to be sure of a huge supply within easy walking distance (you cannot waste good eating time trekking from place to place). The sauropods' long necks gave them access to the higher, tree-top vegetation that other dinosaurs could not reach, and by the end of the Jurassic their herds dominated the landscape. Other new types of dinosaurs evolved alongside the sauropods, including the large meat-eaters like *Allosaurus* and plated plant-eaters like *Stegosaurus*. Smaller coelurosaurs like *Coelurus* and *Ornitholestes* hunted lizards, mammals, frogs and insects among the ferns, horsetails and mosses.

Apart from flying insects, including early ancestors of bees and flies, the air was dominated by the pterosaurs — the flying reptiles from the same original archosaur group as the dinosaurs. *Archaeopteryx* also appeared in the Jurassic Period. This is the earliest known bird and retains many distinctive dinosaur features, proving that birds actually evolved from small meat-eating dinosaurs.

In terms of sheer size and geographical spread, the Jurassic Period is the high point of the dinosaur era. The fully improved stance, so effective in giving the early dinosaurs an evolutionary lead over the other archosaurs, also allowed them to develop varied body shapes and sizes to exploit different parts of the ecosystem. By the end of the Jurassic, dinosaurs had expanded to fill virtually every usable part of the land surface. Increasing numbers of the largest sauropods roamed in herds and trampled large areas of vegetation.

NEW PLANTS. The Earth's plants evolved alongside the dinosaurs and provided food for the plant-eaters.

Late Triassic plants included *Wielandiella*, a bennettitalean shrub closely related to flowering plants.

Ferns like *Gleichenites* were abundant on drier ground during the late Jurassic.

Flowering plants first appeared in the Cretaceous. Forest trees like this grew on drier soils in the late Cretaceous.

THE EARLIEST BIRD. *Archaeopteryx*, the earliest known bird, found in Germany in rocks which are 147 million years old, shows a mixture of reptile and bird features (*opposite*). These include a reptile-like bony tail and claws on the fingers, as well as feathers and a 'wishbone' which are characteristic of birds. The skeletons of *Archaeopteryx* and some small meat-eating dinosaurs show so many similarities that most palaeontologists agree that birds are their descendants. So birds are the dinosaurs' closest living relatives.

THE CRETACEOUS PERIOD

The two supercontinents of Laurasia and Gondwana continued to move apart during the Cretaceous Period (145–65 million years ago), and by the end of that period the outlines of continents were roughly those that we recognize today. The early Cretaceous climate was warm with wet and dry seasons. Later, more pronounced summer and winter seasons developed, although the temperature probably never dropped below freezing even in the polar regions.

The biggest single environmental change during this time was the appearance of flowering plants (angiosperms). It may be that trampling of slow growing plants by large dinosaur herds in the late Jurassic was the key to the rapid spread of the angiosperms, which grow back very rapidly when they are trampled down (ask any gardener). Their seeds would have been easily dispersed by wind, insects and passing dinosaurs. Starting as small herbs, they came to be the dominant plant group, virtually replacing the ferns and horsetails with a dense ground cover of small flowering plants, bushes and trailing stems. They also formed large forests and woodlands of deciduous trees.

The combined effect of the land break up and the new plant life had a direct effect on the development of the dinosaurs throughout the Cretaceous Period. Because there were fewer land bridges, dinosaur populations on each of the continents remained isolated from each other (this had already begun to happen in some areas like Australia). There came to be very noticeable differences between the species developing in each area — the North American hadrosaurs and the equivalent groups in China or Africa are a good example of this. Indeed, in China and Mongolia recent discoveries of strange meat-eaters like *Segnosaurus* and the peculiar looking *Oviraptor* seem to show that entire groups existed here that were not found anywhere else. The discovery of a new meat-eater, *Majungatholus* (described in Chapter 8), from Madagascar, was a real surprise — its closest relatives lived in India and South America. India and Madagascar may have been joined together until late in the Cretaceous Period and there might have been a land-bridge link to South America that bypassed Africa. In Europe, which was covered by a continental sea and divided into large islands, this local specialization took an even more extreme form with the evolution of dwarf versions of larger dinosaurs such as the small iguanodontids *Rhabdodon* and *Craspedodon* and the miniature armoured dinosaur *Struthiosaurus*. Many kinds of animals that we would recognize without difficulty today first appeared during the Cretaceous Period. These included snakes, some birds and moths. The flying reptiles continued to develop and the largest of these were the biggest animals ever to take to the air. *Quetzalcoatlus*, for example, had a great wing span of up to 11 metres, about the same as that of the World War II fighter aeroplane, the Spitfire.

The best exposed area of late Cretaceous fossil-bearing rock is found in western North America. Here the fast-growing flowering plants created great dinosaur ranges that were able to support huge herds of plant-eaters such as *Triceratops*, *Kritosaurus* and *Parasaurolophus*. The bones of up to 10,000 plant-eaters of the same species have been found in one area, giving an indication of just how extensive these herds were. Preying on the plentiful supply of meat were the great carnosaurs such as *Albertosaurus* and *Tyrannosaurus* and smaller scavengers such as *Troodon* and *Dromeosaurus*. These would also have attacked the many smaller plant-eaters in the region. And yet the picture here is only a local one and this pattern of dinosaur life was not repeated elsewhere in the world. Indeed, the development of such different local dinosaur populations makes it difficult to draw many general conclusions about the dinosaur world at this time.

What is certainly true is that we know of more different species from the late Cretaceous than from all the other dinosaur periods put together. This part of the dinosaur era was the highpoint of their development in terms of variety and sheer numbers. The fact that these huge populations of different animals were all wiped out at the end of the Cretaceous remains one of the greatest mysteries of dinosaur science.

GETTING

2

One of the keys to the dinosaurs' success was their way of walking upright on straight legs. This gave them an advantage over other animal groups, which were sprawling

ABOUT

2

walkers, and later allowed them to adopt a wide range of lifestyles. Plodding giants, huge hunters, roaming herds and swift-footed scavengers each evolved their own variation of walking upright.

Dinosaurs walked with straight legs tucked underneath their bodies. It was this ability, never achieved by any other reptile before or since, that opened the way to the evolution of a variety of body types and lifestyles. This helped them to become the dominant land animals and remain so for 160 million years.

Four-legged land animals with a backbone (vertebral column) walk in one of three ways. A lizard sprawls with legs held at right angles to its body. Forward movement is achieved by throwing its weight from side to side and moving its legs one at a time. A crocodile also sprawls most of the time but can sprint for short distances with its legs nearly straight and its body lifted off the ground. This is called the semi-improved stance. All mammals and birds have a fully improved stance, with legs held straight under the body at all times, whether walking, running, galloping or hopping. Humans walk like this and so did dinosaurs.

There are many advantages to the fully improved stance. You do not have to drag your body along the ground to walk like the lizard does, or use huge amounts of energy throwing your weight about. The wrist and ankle joints of sprawlers, and even semi-improved walkers, also have to withstand great twisting and pulling as the animal moves. Nothing very big could walk in this way, putting a severe limitation on the way that such a species can evolve.

Recent research also suggests that sprawling makes it difficult to breathe and move at the same time, the constant body bending interfering with the lungs' action. Fully improved walkers do not have this problem. The dinosaurs achieved what is called 'locomotor stamina', and this ability to run and breathe at the same time ultimately gave them a huge advantage over the sprawlers.

Having a fully improved stance allows an animal to grow bigger, walk further and move faster, and this opens the evolutionary path to a huge range of body types and lifestyles. This can be clearly seen with today's fully improved walkers, birds and mammals — some are huge, some fast, some tall, some agile. The dinosaurs too, having once achieved fully improved walking, evolved to exploit it in many different ways.

A PLODDING GIANT

The large sauropods like *Diplodocus* used the fully improved stance to develop to a huge size, so large that a fully grown adult could not be attacked by any predator. Some remarkable physical adaptations were required to achieve such bulk. *Diplodocus* stood on four thick, widely spaced legs. These supported its huge shoulder bones and hip girdle, rather like pillars holding up the cross beams of a building. The five vertebrae across the hip were actually fused together to form a strengthened support for a hip girdle that could take the immense pressure of an 11 tonne body on the move. All four legs ended in short, broad feet rather like an elephant's. *Diplodocus*' front toes were probably rounded horny pads, but the inner toe on the front foot had a much longer, sharper claw. There were sharp claws on the three inner toes of the back foot as well and these were probably an anti-slip device, able to dig into soft ground and prevent *Diplodocus* from skidding or getting stuck at the bottom of a slope.

Diplodocus and other sauropods also had built-in 'high heels'. When you walk you raise your ankle at each stride to lift your whole body weight. This is actually quite tiring and the reason why shoes with heels make walking easier. If lifting a 68 kilogramme human an inch above the ground at each stride is hard work, imagine the energy cost to an 11 tonne *Diplodocus*. Elephants have the same problem today, and have solved it by growing a thick

(*Previous spread*) A GIANT AFLOAT. One set of dinosaur trackways seems to show *Apatosaurus* walking on its front feet only! Scientists have concluded that it was floating in water, pushing itself along with its front feet and steering with the back legs.

THREE WAYS TO WALK. Four-legged animals with a backbone walk in one of three ways.

Monitor lizard. Sprawling, with legs at right angles to the body. The early ancestors of dinosaurs walked like this.

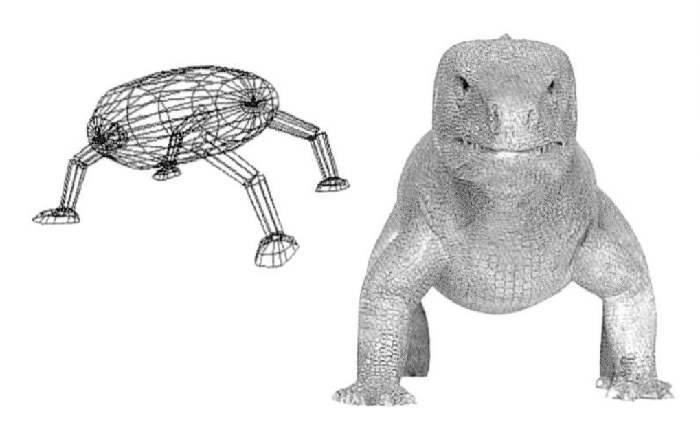

Euparkeria. Semi-improved, with legs straighter and body held higher off the ground. This can only be maintained over short distances.

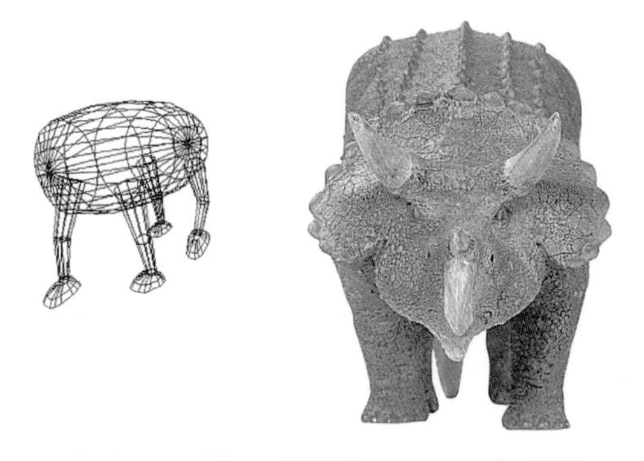

Triceratops. Fully improved, with straight legs tucked under the body. The upright stance was the key to the dinosaurs' success.

wedge of fibrous tissue under the heel of each foot. The same solution worked equally well for the sauropods.

Although quite a few sauropod trackways have been found around the world, there are very few that show evidence of tail-drag marks. We can only conclude that *Diplodocus* and the other sauropods kept their tails up off the ground when they walked — a sensible arrangement, since a heavy tail would soon get worn down to the bone after a few kilometres on the move. This would have been achieved by large hawser-like ligaments which ran along the tail and attached to the top of the hips. Similar ligaments ran all the way down the sauropods' long necks, through V-shaped guide tracks in the tops of the spines of the back vertebrae, also attaching to the hip area. These two sets of ligaments worked together, allowing the weight of the neck and tail to balance each other and keeping both off the ground.

Being on the move was not a speedy business for this group of dinosaurs. As far as we can tell from their physical structure and the trackways they left behind, most sauropods moved at about 6 km/h — human walking pace. Perhaps, like elephants, they were capable of modest bursts of speed, but the forces of torque and compression on their bones and tendons would have been hugely increased by even a gentle trot, so they must have conducted most of their lives at a gentle amble. Sauropod tracks up to 147 metres long, the longest on record, were discovered in Portugal in 1994. The animals that walked over a muddy estuary some 170 million years earlier left very distinctive footprints. The front footprints are much larger than the hind ones and, unlike other sauropod tracks, they show distinctive claw marks on the inside edge of each print. This proves that the large thumb claws of sauropods were in contact with the ground each time the animal put its foot down.

A WALK BESIDE THE SEASIDE. This 147 metre-long sauropod trackway from a site in Portugal is the longest trackway on record. The track was made by an animal walking in a straight line across a muddy estuary.

Argentinosaurus, recently found in Argentina, will probably will turn out to be the largest dinosaur yet discovered. The dorsal vertebrae are almost 1.65 metres tall but they were lightly built, with weight-saving features as described in Chapter 6. Powerfully developed extra joints between adjacent vertebrae helped to stiffen and strengthen the spinal column, to support the creature's tremendous weight. It was about 27.5 metres long (about the same length as *Diplodocus*) and about 6 metres tall at the shoulder. Its length was mostly accounted for by neck and tail; its body was not much larger than a big African elephant. *Argentinosaurus*, with a bulky body 50% longer, taller, and wider than *Diplodocus*, probably weighed almost 3.5 times as much, up to 50 tonnes. The record for the longest dinosaur is currently held by *Seismosaurus*, a *Diplodocus* relative, estimated at between 39 and 52 metres.

The extraordinary size of the sauropods gave rise to some bizarre theories, in the early days of dinosaur study, about how they moved. Until quite recently some scientists believed that an animal as huge as *Brachiosaurus* could never have

VARIATIONS ON A THEME. Today's land animals show how versatile the improved stance can be, supporting many different body types and methods of moving. The elephant is a heavy, slow moving quadruped; the ostrich is a fast bipedal runner; the cheetah runs very fast on all four legs; the polar bear can walk on two legs as well as on all fours. Dinosaurs exploited the improved stance in similar ways.

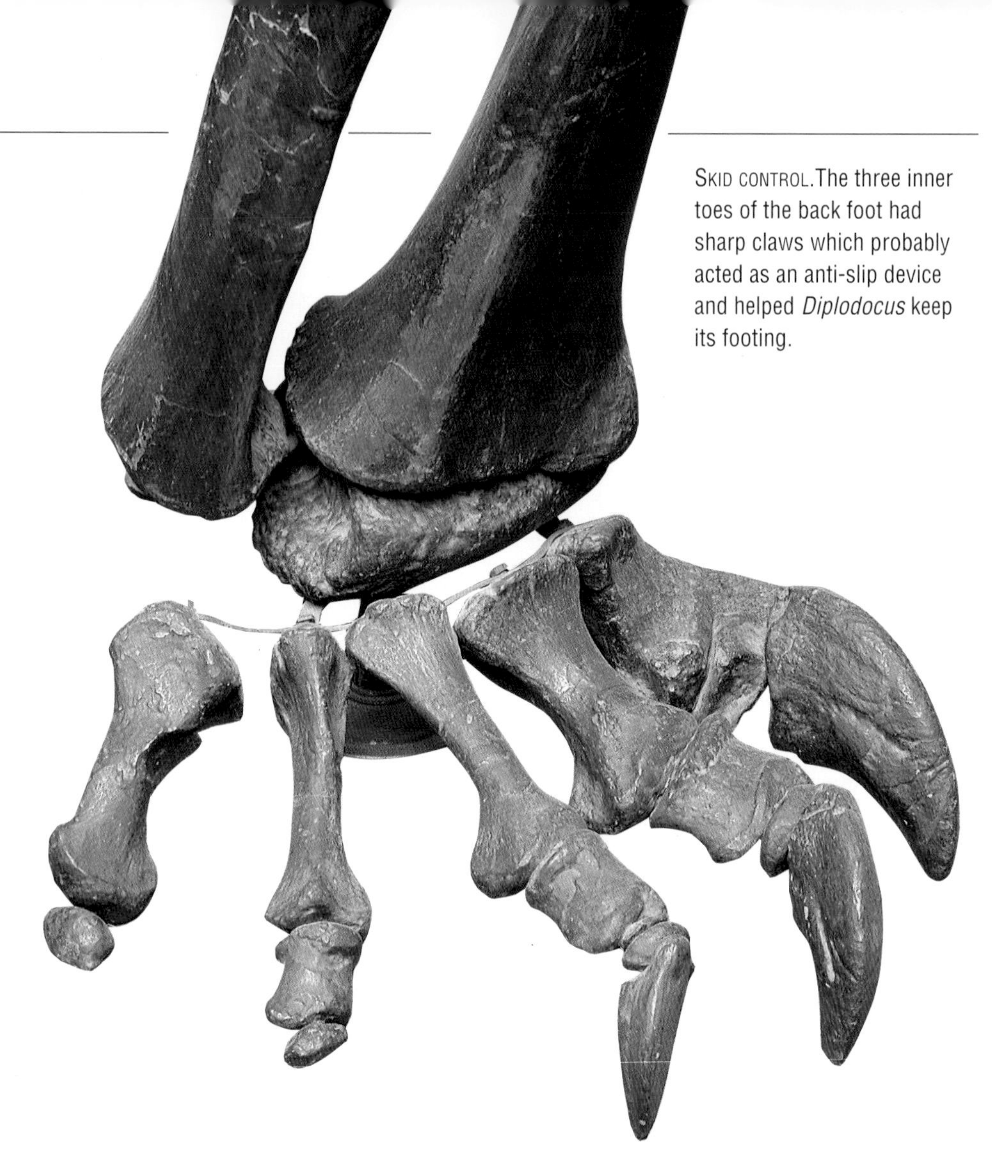

SKID CONTROL. The three inner toes of the back foot had sharp claws which probably acted as an anti-slip device and helped *Diplodocus* keep its footing.

BUILT FOR SIZE. *Diplodocus*' skeleton shows how the improved stance was used to support huge weight. The legs were like pillars holding up the cross beams of shoulders and hips. Ligaments ran from one end of the spine to the other keeping the head and tail up without wasting energy.

supported its own body weight on land and must have lived in the water. The fact that all sauropods had nostrils on the tops of their heads was seen as further evidence of this. In fact, *Brachiosaurus* had a physique that was quite capable of supporting itself on land; the pressure exerted on its lungs if it had been submerged in water would have meant instant death. Even earlier, when dinosaurs were still thought of as huge lizards, the sauropods were said by some German scientists to have had sprawling legs and to have rested their huge bodies on the ground. The discoverers of the original *Diplodocus* in America, however, showed that the dinosaur would have needed a 2 metre deep pit underneath its belly to stand in this position, and a network of trench runways to move about! These early mistakes are an indication of just how extraordinarily different the sauropods were from any living animal and how difficult scientists found it to understand them. Today we can see the sauropods as the supreme example of what the improved stance made possible for dinosaurs in terms of sheer bulk on an evolutionary path that sacrificed everything to size.

A DINOSAUR SPRINTER

At the other end of the scale, *Hypsilophodon* shows how the fully improved stance allowed the development of impressive speed and agility. *Hypsilophodon* was a two-legged plant-eater standing just 1.5 metres high. Many well-preserved skeletons have been found in one location on the Isle of Wight, England, where at least twenty-five *Hypsilophodon* died together, possibly after wandering into an area of quicksand. The most striking feature about a *Hypsilophodon* skeleton is how little there is of it. Rather like a gazelle or antelope, the whole structure had been slimmed down to give maximum support for minimum weight. Even the bones themselves were thin walled and hollow, just like a gazelle's. The thigh bone of *Hypsilophodon* is quite short, allowing it to be pulled rapidly back and forth for fast strides. The foot, by contrast, is remarkably long and thin, with very elongated upper foot bones (metatarsals), and thin toes ending in short, sharp claws. This structure gave a very flexible grip on the ground and could sustain long periods of fast running. The completely defenceless *Hypsilophodon* must have made frequent use of its speed to escape its enemies, helped by its long tail, stiffened by bony rods. With *Hypsilophodon* in its typical horizontal running posture, the tail could have been swung from side to side for rapid changes of direction without stopping.

EARLY THEORIES. *Top:* Early dinosaur scientists thought that the largest dinosaurs like *Brachiosaurus* could only have moved when buoyed up by water. *Bottom:* The idea of a sprawling *Diplodocus* was also popular at one time. But how could it have moved with its belly hanging below ground level?

Like the sauropods, *Hypsilophodon*'s way of moving was considerably misunderstood when it was first discovered. Scientists assumed that the big toe on its foot actually faced the opposite way from the other toes, rather like a bird's. *Hypsilophodon*, they said, must have been a tree dweller, perching in the branches and using its tail to balance, just as the Australian tree kangaroo does. In fact, as we now know, *Hypsilophodon* shows how the fully improved stance enabled quite small, unarmoured dinosaurs to thrive by developing high levels of speed and manoeuvrability to escape their enemies.

TWO OR FOUR LEGS

While *Diplodocus* was a four-legged walker and *Hypsilophodon* a two-legged one, there were some dinosaurs that could move in both ways, rather like modern bears. Using two and four legs gave these dinosaurs several advantages. They could rear up and use their hands for gripping food or doing battle with an

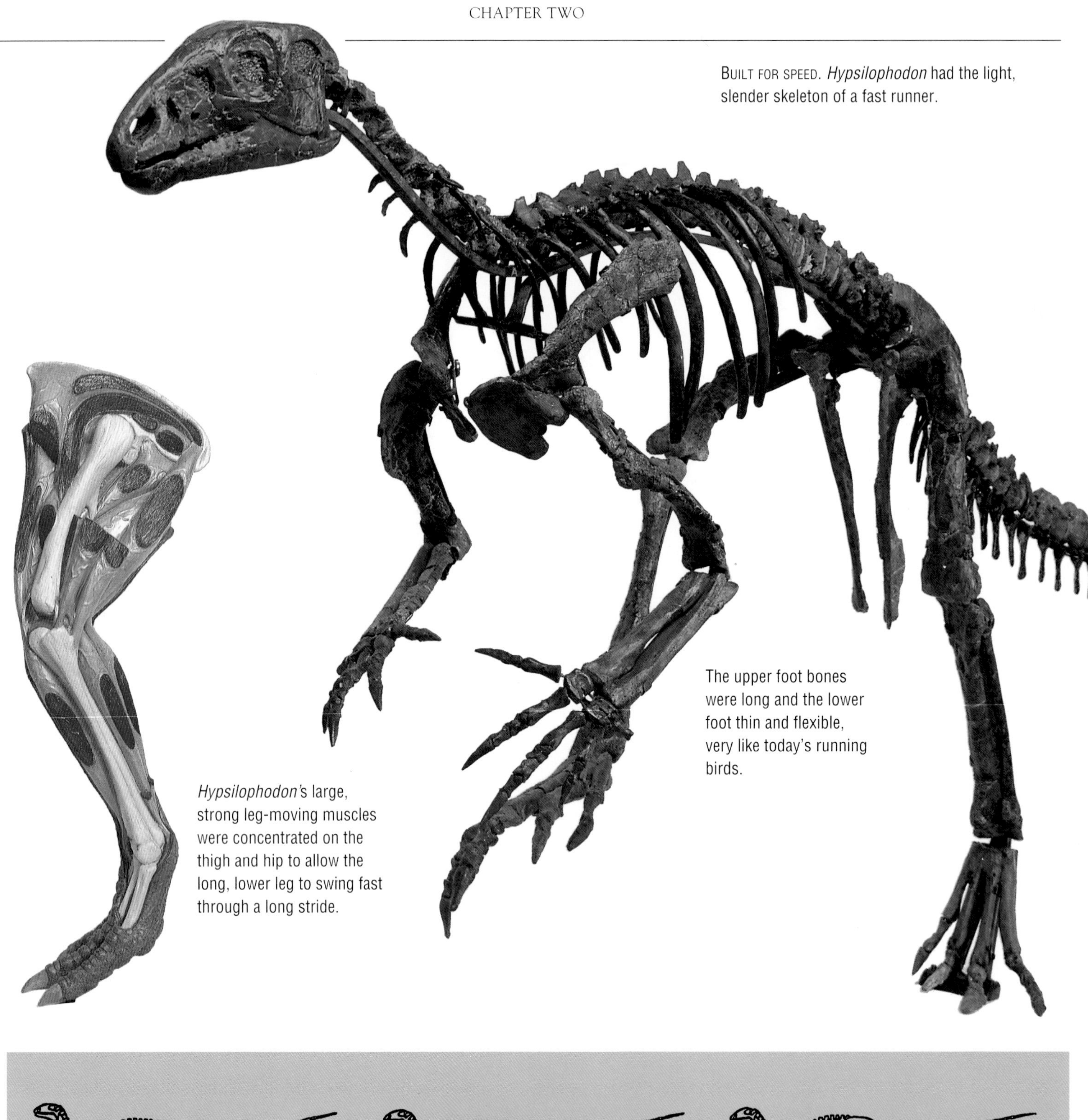

BUILT FOR SPEED. *Hypsilophodon* had the light, slender skeleton of a fast runner.

The upper foot bones were long and the lower foot thin and flexible, very like today's running birds.

Hypsilophodon's large, strong leg-moving muscles were concentrated on the thigh and hip to allow the long, lower leg to swing fast through a long stride.

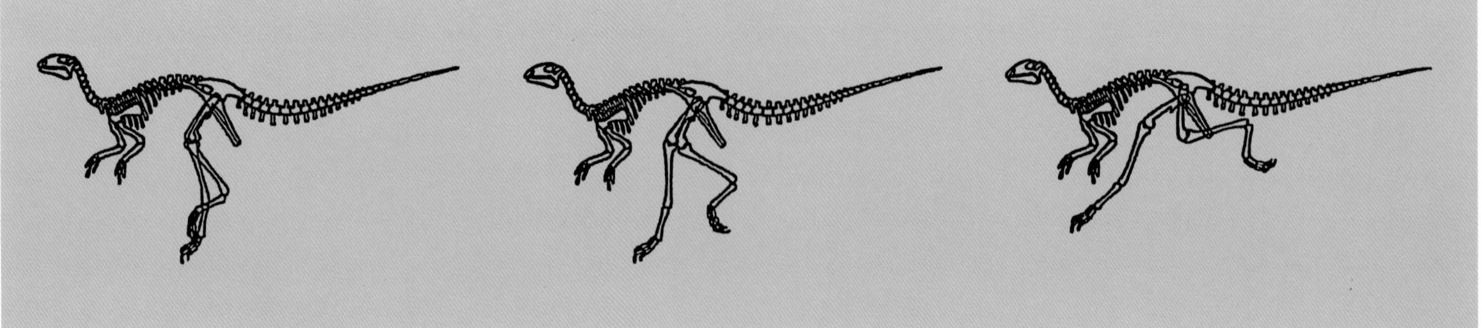

STRIDING OUT. This computer generated picture shows the stride patterns of *Hypsilophodon*, a fast runner with a long stride.

enemy, for example, as well as browsing for low-level food on all fours. They could rest or walk slowly about with all four legs on the ground, but if a sudden increase of speed was required, they could hoist up on two legs and get away.

Iguanodon is one of the best known examples of a dinosaur that exploited the fully improved stance in this way. Related quite closely to *Hypsilophodon*, *Iguanodon* was built on a much larger scale with a fully grown adult reaching up to 10 metres in length and weighing 4 tonnes. Its skeleton had exactly the same basic structure as *Hypsilophodon* but the proportions of the bones were very different. The thigh bone was heavy and long and the upper foot quite short, giving great strength to carry the weight of the body, but no real ability to run. The spines on the vertebrae were much wider and taller, with numerous bony tendons criss-crossing along their whole length to give added strength without the weight of extra muscles. Since *Iguanodon*'s first discovery there has been much debate about just how it held its body in the normal walking stance. Horizontal like a lizard? Upright like a kangaroo? Most scientists now agree that a fully grown adult *Iguanodon* probably walked with a nearly horizontal spine, allowing the back legs to take most of the weight but dropping down frequently on to all fours for extra support when feeding or standing around.

Iguanodon's arms were one of its most remarkable features and show again how very adaptable the improved stance was for the dinosaurs. Supported by huge shoulder bones, the arms were long, powerful and heavily muscled. The five bones of its wrist (carpals) were fused together to give a strong inflexible support, very different from *Hypsilophodon*'s sliding wrist bones. The three middle fingers were strong and stiff, ending in short, blunt claws which

A PERCHING DINOSAUR? *Hypsilophodon*'s foot skeleton was once thought to show that this dinosaur was a tree dweller, similar to the Australian tree kangaroo (*above*). *Right:* This model shows the mistaken idea.

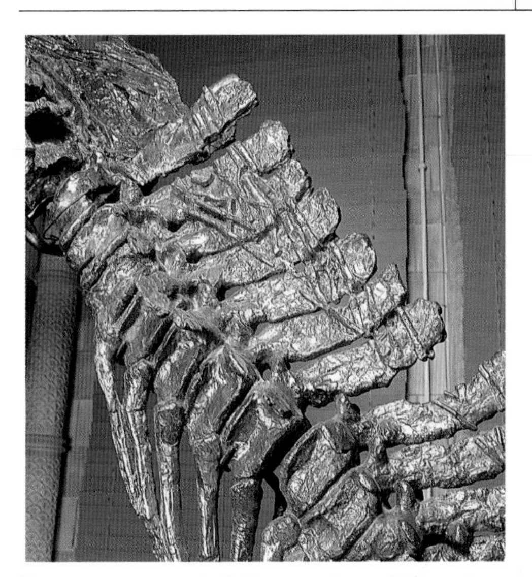

TWO AND FOUR LEGS. *Left:* Bony tendons all along the spine gave *Iguanodon* strength and support. *Right: Iguanodon* was able to walk upright and on all fours.

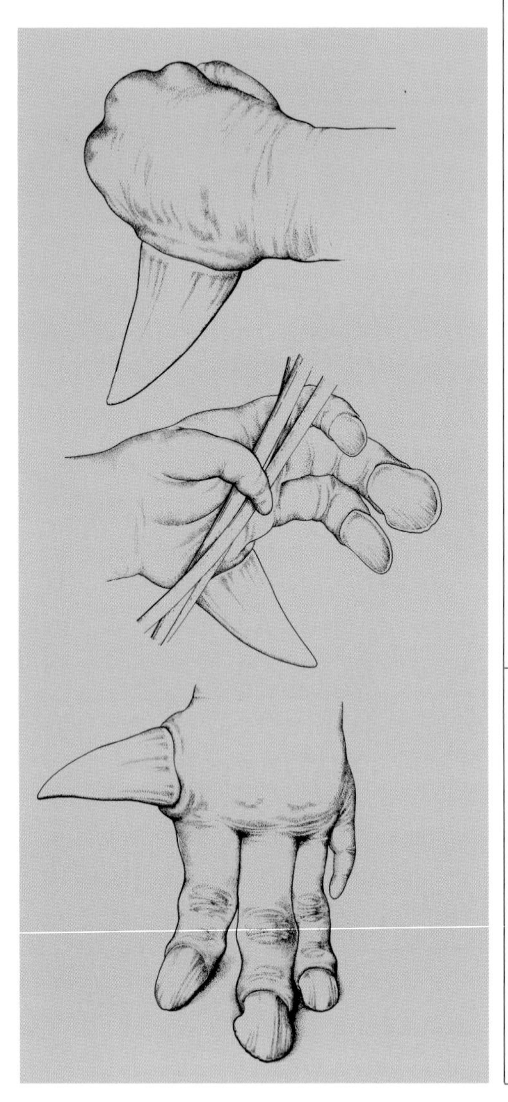

would have splayed out like hooves during four-legged walking. The thumb was a fearsome spike, *Iguanodon*'s main weapon when it reared up on its hind legs in defence. The fifth finger was weaker than the others but much more flexible, and could have been used as a hook to pull down food from the trees.

THE BIG HUNTERS

Of all the dinosaurs it has been the large meat-eaters like *Albertosaurus* and *Tyrannosaurus* whose means of getting about has caused most dispute among scientists. There is general agreement that they had the fully improved stance, that they were two-legged and grew to a great size (up to 7 tonnes for an adult *Tyrannosaurus*). Beyond this point agreement ends, with one group seeing them as slow and lumbering scavengers and another as speedy hunters. Both groups often interpret the same evidence to reach quite different conclusions, and the whole subject is a good reminder of how little we actually know for certain about many aspects of dinosaurs.

As with all dinosaurs, study begins with the skeleton. In the case of *Tyrannosaurus* this was huge and heavy with particularly massive vertebrae, hip girdle and thigh bones to anchor the muscles and tendons that controlled movement. The metatarsal bones of the upper foot were actually locked together to make an extraordinarily strong pillar of bone. The three toes that took its weight were short and powerful, ending in large claws. At first sight this does not look like the body of a fast runner. The slow movement group say that the sheer

ADVANCED HANDLING. *Left: Iguanodon* had an extraordinarily multi-purpose hand. The thumb carried a large spike for self-defence while the three middle fingers would have splayed out like hooves for four-legged walking. The flexible fifth finger could grasp food and strip leaves from the trees.

THUMB SPIKE. *Iguanodon* rears up to lash back at an attacking allosaur with its thumb spike. The fully improved stance allowed *Iguanodon* great freedom of movement.

weight of flesh and bone would have prevented sprinting, and that the stresses imposed on the bones by running would have been fatal to an animal of this size. The advocates of fast movement disagree, suggesting that the bulk of muscle and bones would only have developed if *Tyrannosaurus* had used them for running. They point out that some bones are capable of taking the same stresses as mild steel, around 3000 kg/cm^2, and that large modern animals like the rhinoceros can get up to speeds of 35 km/h over short distances.

The theropods' knees are another area that both groups believe prove their own case. Dinosaurs' knee joints were similar to yours, a roller-like hinge shaped to allow forward and backward movement without twisting. In modern mammals the bone ends are highly ossified (bony) with a very thin cap of softer cartilage to allow smooth movement of the joint. Dinosaurs, on the other hand, show evidence of having had pads of much thicker cartilage between the ends of their upper and lower limb bones. This is similar to today's juvenile birds, as you can clearly see should you have a chicken for dinner. This cartilage, say the slow moving enthusiasts, would have been unable to take the shock of a running theropod and was designed to allow the animal to move at a slow plod. Not so, say the fast movers. Cartilage is an ideal material for fast running since it spreads the weight over the whole joint and provides a smooth lubrication for the knee as it moves back and forth.

Trackways are another source of controversy. Literally thousands of meat-eater trackways have been found around the world, some belonging to the largest species. They demonstrate, among other things, that these large meat-eaters seldom travelled in groups, that they walked almost on tiptoe and probably had a slightly bow-legged stance to keep their feet directly underneath them without their knees hitting their own large bellies at each stride. What they do not prove is whether they could run, because all the trackways so far discovered have been of walking animals. This is because they never ran, say the slow moving advocates. The fast moving group point out that most modern animals, including some of the fastest,

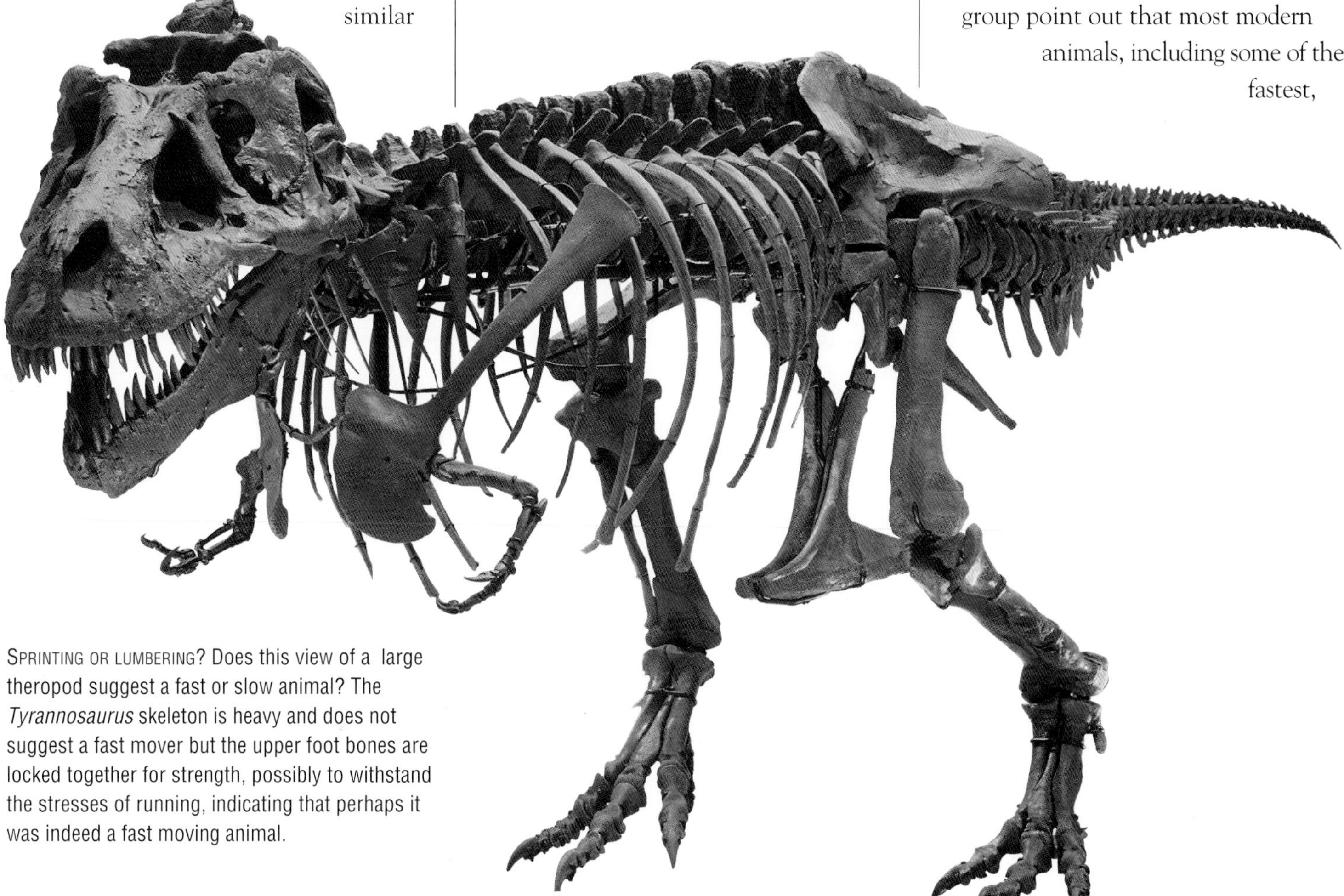

SPRINTING OR LUMBERING? Does this view of a large theropod suggest a fast or slow animal? The *Tyrannosaurus* skeleton is heavy and does not suggest a fast mover but the upper foot bones are locked together for strength, possibly to withstand the stresses of running, indicating that perhaps it was indeed a fast moving animal.

spend most of their lives walking and that running trackways are therefore very rare. However, the fact that none have been found does not mean they do not exist.

For the moment, with the direct evidence still in dispute, our conclusions about the big theropod's movement must rest on deduction. *Tyrannosaurus* was a huge meat-eater and hence required huge quantities of meat. It would be unlikely to find this lying around as carrion, so it must also have hunted. Unless it was a master of ambush and disguise it must have been able to move at least as fast as its prey, outrunning hadrosaurs like

TRIPPING *TYRANNOSAURUS.* If a 6 tonne *Tyrannosaurus* tripped and fell at top speed it would probably have been killed outright. Rocks and fallen branches would have been more of a hazard than tiny mammals!

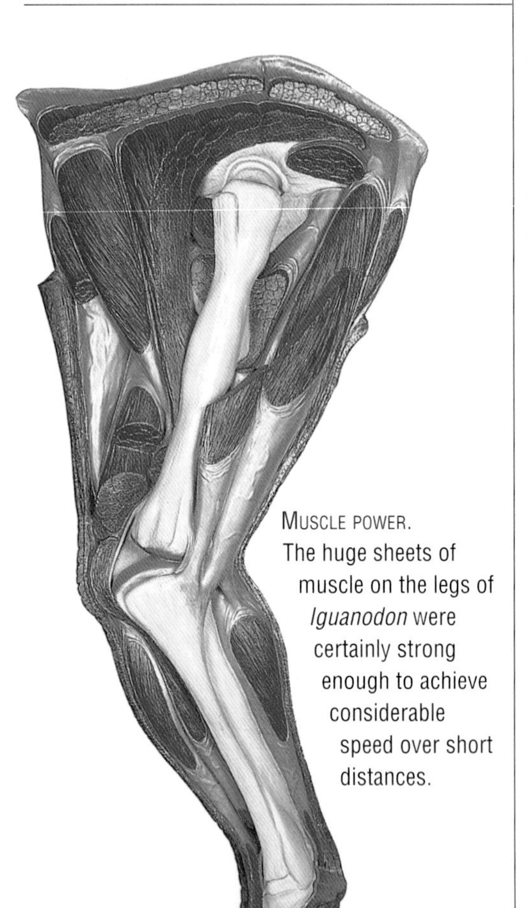

MUSCLE POWER. The huge sheets of muscle on the legs of *Iguanodon* were certainly strong enough to achieve considerable speed over short distances.

Edmontosaurus and ceratopians like *Triceratops*. Some of these dinosaurs are believed to have been able to reach 14–20 km/h at top speed (and running away from a large meat-eater would definitely justify top speed), so *Tyrannosaurus* may have been able to do the same over very short distances. Running speeds in *Tyrannosaurus* have been subjected to complex mathematical calculations based on a very complete skeleton in the Museum of the Rockies, Montana. The results suggest that its safe top speed was about 10 m/sec, or 36 km/h, over a short sprint, more than fast enough to catch its prey. There have been other suggestions that *Tyrannosaurus* could have run much faster than that. However, the calculations indicate that if a 6 tonne *Tyrannosaurus* were to fall while running at 20 m/s, the deceleration and impact forces would kill it outright. Its huge head could have hit the ground with a force of up to 16*G*. Running that fast, even if theoretically possible, would place enormous stresses on limb bones and joints too. It is unlikely that any animal would risk even a non-lethal injury that would compromise its essential daily activities. *Tyrannosaurus* was probably very careful about when and where it conducted its sprinting activities. In contrast, based on its tracks, the small meat-eater *Scartopus* might have been moving at a comparatively leisurely 16 km/h calculated from the size and spacing of the footprints it left in Lark Quarry in Australia.

The large meat-eaters show yet another adaptation of the fully improved stance, combining bulk with a hunter's speed and power. So successful was this adaptation, in fact, that they became the largest hunters ever to live on Earth, a clear testimony to the effectiveness of the fully improved stance as an evolutionary pathway.

FOOTPRINTS IN THE ROCK

Trackways are obviously important in understanding dinosaur movement but occasionally a set of trackways can reveal much more information, even a complete incident from the prehistoric world. One such set has been found at Lark Quarry in Australia and is thought to show a dinosaur stampede.

HUNTING REPTILE. Usually a slow mover, the Komodo dragon can run at up to 14–18 km/h in pursuit of prey. The big theropod dinosaurs must also have been able to move at least as fast as their prey.

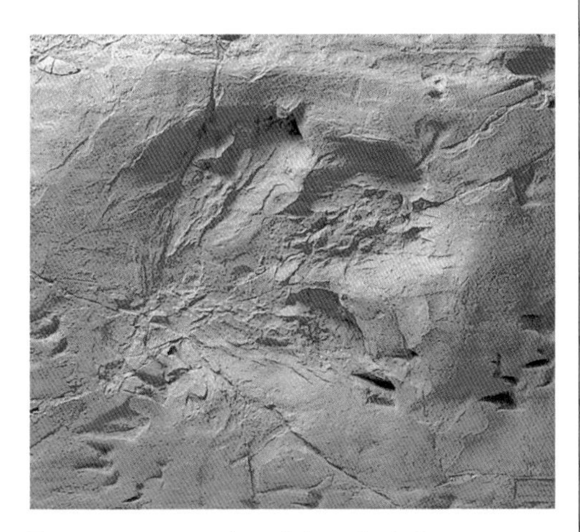

TRACKWAY EVIDENCE. A track from Lark Quarry showing the smaller dinosaurs' footprints covering that of the big hunter.

The trackways date from the Lower Cretaceous Period and probably form part of the dried up bed of a small stream or river that was still muddy enough to take an impression of footprints. The preserved section covers 209 m^2, running from north to south, and probably led down to a waterhole, all traces of which have now disappeared. From the trackway evidence we get a picture of about 150 small meat-eating and plant-eating dinosaurs gathered around this waterhole. Presumably the plant-eaters were there in sufficient numbers to prevent an attack by the meat-eaters, but both groups must have been keeping a wary eye on each other and on the surrounding countryside.

From the northern end of the watercourse there appears a large meat-eating dinosaur, standing 2.6 metres at the hip, which advances four paces. The footprints then suggest a change of tempo, almost as if it was here that the hunter spotted the group of smaller prey. The stride becomes shorter and the foot pad impressions disappear as it moves forward on tiptoe for five further steps and then turns. Some time in these final few seconds the small dinosaurs at the waterhole see the hunter. There is instant panic as all of them stampede back towards it. Why towards? It is impossible to say. Perhaps the waterhole was, in fact, a wide lake that cut off their retreat. Perhaps another meat-eater was stalking them from the opposite direction. In any case, the whole group (less any that the hunter picked off on the way past) charged back up the watercourse, covering it with the footprints that we can see today.

Dinosaur locomotion is important because it was vital to their success. In the early stages of their evolution it was the only thing that distinguished them from the many other animals that could have become dominant on land. Later on, its energy efficiency and above all its versatility set the dinosaurs on an evolutionary track more spectacularly diverse and successful than any group of land animals had ever achieved before. Dinosaur diversity has been surpassed only by that of the mammals which eventually replaced them. Mammals include burrowers, fliers and swimmers — dinosaurs never went underground, took to the air or became aquatic. However, one group of dinosaurs did learn to fly — the birds. As we saw in Chapter 1, *Archaeopteryx* is really just a flying dinosaur, and, as we shall see in Chapter 8, recent fossil discoveries chart the stages in the transition from small meat-eating dinosaurs through to modern birds. These changes began first with the forearms evolving into feathered wings. The hip, hind legs and tail were modified later, leading to major changes in body balance, locomotion, and eventually the loss of a long bony tail. So dinosaur locomotion was drastically remodelled to suit a very different and very successful lifestyle in the air.

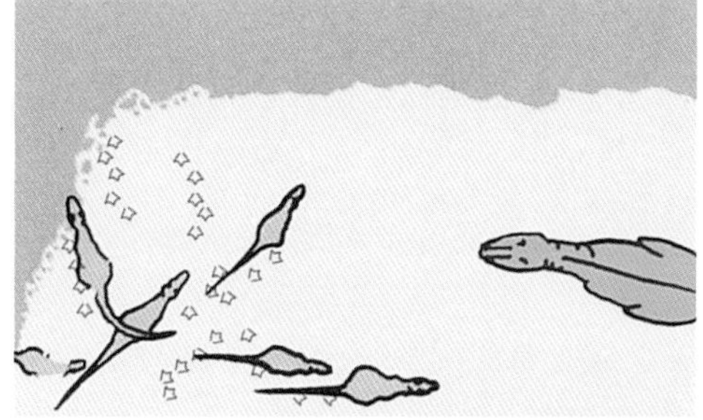

STAMPEDE! This computer reconstruction shows the Lark Quarry trackways and the possible sequence of events. *Top left:* Smaller dinosaurs gather at the water course. *Top right:* Large meat-eater approaches from northern end. *Bottom left:* Smaller dinosaurs stampede back past the hunter. *Bottom right:* The hunter closes in on escaping small dinosaurs, and a mass of large and small trackways are left behind.

FOOD AN

3

As with all today's animals, the base of the dinosaur 'food chain' was plants and the species that ate them. These provided meat for the hunters and scavengers that preyed on

D DRINK

herds or individual victims. Some dinosaurs also developed remarkable specialist equipment, such as grasping hands, swivelling claws and grinding teeth, to catch and process their food.

Animals either eat plants or each other. A plant-eater's food is not hard to catch (being literally rooted to the spot) but can be very hard work to process and digest. This is because plants are largely composed of cellulose, a chemical compound that is difficult to break down. If a plant-eater is to get enough nutrition, it needs a complex food processing system. Meat-eaters have a much easier task getting energy from their food. Animal flesh and fat are both energy rich and break down rapidly under the action of digestive acids and enzymes. The meat-eaters' problems lie in catching their food, which tends to run away or fight back, and eating it quickly before it turns rotten and poisonous.

Most dinosaurs can be divided clearly into plant-eaters and meat-eaters. The equipment they used to obtain and process their food is well known as far as jaws and teeth are concerned, and we can tell fairly accurately from the fossil record what there was around to be eaten (see Chapter 1). Beyond these facts there is much less certainty. After all, it is precisely soft internal organs like the stomach and gut that are the first to rot away when an animal dies, and no part of a dinosaur's digestive system has ever been found as a fossil. We must rely instead on deduction from the fossil evidence and the application of modern animal dynamics.

A TONNE A DAY

The biggest dinosaurs of all were the plant-eating sauropods of the Jurassic Period. To obtain enough nutritional energy to keep their bodies going, these giants must have consumed vast quantities of plants. The nearest modern equivalent would be the African elephant, which must eat around 185 kilogrammes of plants every day to survive, over 3 per cent of its own body weight. At this rate, a 30 tonne *Brachiosaurus* might have chomped through around a tonne of vegetation per day. Yet all this consumption was achieved with a head less than 75 centimetres long, not very much larger than a horse's, and teeth which did no chewing at all.

For many years the sauropods' small heads and teeth convinced scientists that they could only have survived on a diet of aquatic plants. *Diplodocus*, for example, had just a short row of pencil-thin teeth at the front of its mouth. With such equipment, it was said, only the softest water-weed soup could

CHOP AND RAKE. *Left:* These robust chopping teeth from *Pelorosaurus,* a sauropod, show where the outer enamel has been worn away by constant chopping at plants.

(Previous spread) A HUGE HUNTER. A lone-hunting *Allosaurus* makes a savage assault on a young *Diplodocus*. The victim will provide food for several days.

possibly have been processed in sufficient quantities to keep it alive. Yet why should the sauropods develop such extremely long necks, up to 11 metres in the case of *Mamenchisaurus*, if they were only going to use them for water-level browsing? Computer reconstructions of neck posture and range of movements in *Diplodocus* and *Apatosaurus* show that their necks were nearly straight. In fact they sloped gently downwards, so that their heads, which were angled downwards too, were close to the ground in a neutral, relaxed pose. Both necks were less flexible than conventionally depicted, and *Diplodocus* was less capable of sideways and upward curvature than *Apatosaurus*. The results suggest that these sauropods were adapted to ground feeding or low browsing and their long necks enabled them to gather plants well before these were trampled underfoot. It does seem likely that other sauropods like *Brachiosaurus*, which has long front legs, were mainly tree-top browsers and their long necks were for reaching high trees where, as any giraffe will tell you, the leaves are tender and no other animal can get at them.

Yet even if sauropods' lives were one long meal among the tree tops, how were such huge quantities of food processed, particularly in the absence of chewing teeth? The answer lies in the rounded pebbles that are occasionally found at sauropod fossil sites. These pebbles (called gastroliths) are polished, smooth and made of quite different material from the surrounding rocks. They come from the gizzards of sauropods and were a vital part of their eating equipment.

The best modern comparison is a bird's gizzard. A strong muscular sack, the gizzard is the second part of a bird's stomach which it fills by swallowing grit or small stones. Food passes through the bird's first, glandular stomach into the gizzard where muscular action grinds the stones together like a gastric mill, reducing the food to a thick paste. This is then sent on to be further broken down by bacteria and the nutrition extracted. A gizzard is a very good way of dealing with all kinds of tough material. Ducks can crush clam shells with theirs, while the tropical fruit pigeon's gizzard cracks the toughest nuts and seeds. This is how the big sauropods managed to process their daily tonne of plants, saving themselves the bother of chewing. Their gizzards, perhaps similar in structure to those of some birds, would have worked effectively on anything they found in their tree-top browsing, and certainly not restricted them to a diet of soft, water-plants.

If today's animals are anything to go by, dinosaurs may well have been very choosy about where they obtained their gastroliths. Many birds prefer the hardest rocks available, such as quartz which wears down slowly, and will travel for long distances to find the right sort. Some of the gastroliths from dinosaur fossil sites also originated many kilometres from where the dinosaur actually died. One *Massospondylus* skeleton found in Zimbabwe had gastroliths that came from at least 20 kilometres away.

It is certain that the big dinosaurs would have got through a lot of gastroliths in the course of a year's eating. A specimen of the North American *Barosaurus* was found with sixty-four polished stones between its ribs. When a bird's gastroliths become too smooth to work properly it belches them up and takes new ones on board. It is entertaining, although probably fanciful, to speculate about a peaceful dinosaur landscape rent asunder by whizzing gastroliths regurgitated at high velocity by hungry sauropods.

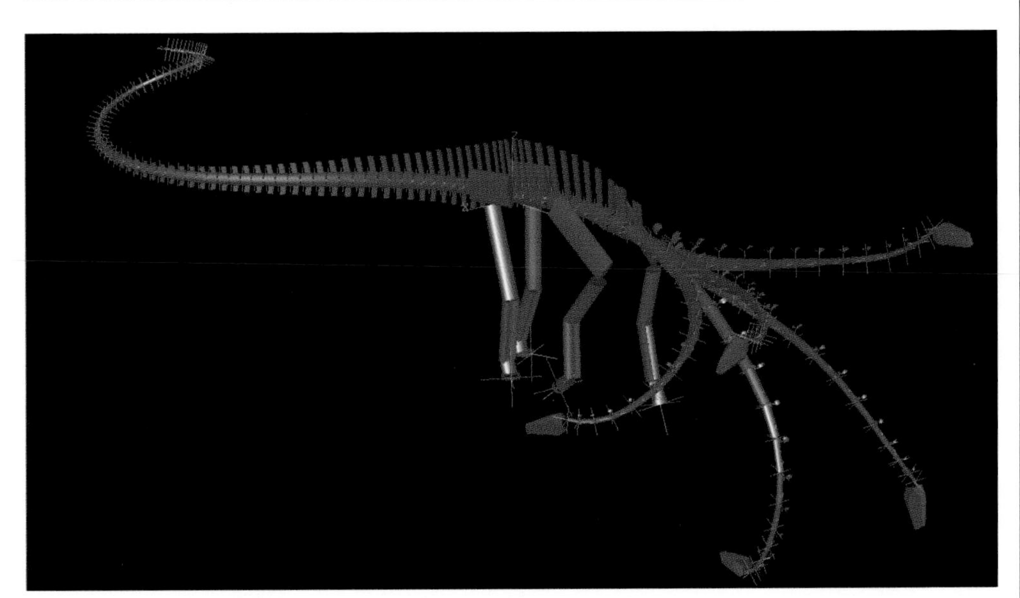

CRANING THE NECK. A computer-generated model of *Diplodocus* shows the range of possible neck movements. The extreme high position to the right and low positions suggest that *Diplodocus* was a low browser or ground feeder that could not reach the tree tops.

CHOPPING AND GRINDING

For plant-eating dinosaurs without

STONE CRUSHERS. Modern birds have gizzards, a muscular part of the stomach in which small stones or grit crush up the food before it reaches the intestines. Dinosaur stomach stones (gastroliths) have been found near, and sometimes actually inside, the ribs of sauropods and parrot dinosaurs.

gizzards it was teeth and jaws that had to crush and shred the food for the digestive system, just as with modern grazing and browsing mammals. All these dinosaurs had grinding or chewing teeth, but two in particular demonstrate how effectively different groups adapted to take advantage of particular kinds of food.

Edmontosaurus was a herding hadrosaur from the Upper Cretaceous Period. By this stage the dominant ferns, cycads and horsetails of the earlier dinosaur periods had been overtaken by the flowering trees and bushes. Tough leafed conifers of various kinds were also common. *Edmontosaurus* had a wide, flat beak covered in horn that cut or pulled at the vegetation. Its powerful tongue would then push the food back towards the multiple rows of cheek teeth, called tooth batteries, where an extraordinary grating and grinding process took place. The upper teeth fitted neatly over the outside of the lower teeth to give two interlocking grinding surfaces. The teeth were constantly replaced by new ones growing up underneath so the tooth battery was always being renewed. On their own these would have been impressive chewing equipment, but there were a number of additional refinements that made *Edmontosaurus* a particularly efficient eater.

First, it had cheeks as did almost all other plant-eating dinosaurs. This may seem a small achievement, but no modern reptile has them, which makes them relatively inefficient eaters. When a tortoise chops a blade of grass, at least half falls from each side of its mouth on to the ground. *Edmontosaurus*' efficient chewing would have been useless without muscular cheeks to hold the food in place. The second feature was the rough, grater-like surfaces of the tooth batteries. This developed because the hard outer enamel of each tooth wore away more slowly than the softer core of dentine, giving a pattern of dips and ridges that could shred the toughest greenery. Finally, the whole upper jaw of *Edmontosaurus* was specially hinged for chewing. When humans chew, our lower jaws swing slightly from side to side with each bite to let the back molars grind against each other. In *Edmontosaurus* the upper jaw moved slightly outwards on both sides, hinged by great sheets of muscle just below the eyes. The pressure of the closing lower jaw pushed the upper jaws out sideways so that the upper and lower tooth batteries could shred and crush at the same time. Analysis of the stomach contents of exceptionally well-preserved fossils shows that hadrosaurs ate, among other plant material, bark, pine-cones, branches and the toughest kind of conifer needles — food that no other dinosaurs could handle. As with *Brachiosaurus* in the high branches, this left *Edmontosaurus* and its relatives as the unchallenged owners of a specialized area of food supply.

TOOTH BATTERY. Two views of the teeth in the lower jaw of a plant-eating hadrosaur. There are more than 200 teeth in each jaw.

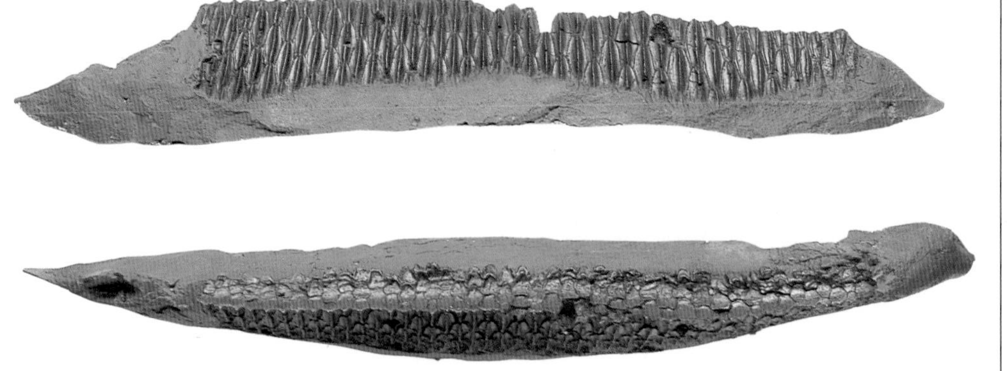

The teeth worked together as a single grater-like surface, which ground up the toughest plants.

A very different kind of food specialization was developed by *Protoceratops* and the other horned dinosaurs. At the front of their mouths a sharp, narrow beak sliced off the chosen shoots and leaves. Further back the scissor-like tooth batteries operated a fine cutting action, chopping but not grinding the food before it was swallowed. This system was well suited to dealing with tough, low-growing fibrous plants such as palms and cycads which *Protoceratops* may have eaten.

Brachiosaurus, *Edmontosaurus* and *Protoceratops* show three different strategies that plant-eaters can adopt to compete effectively for food — special access, special processing and specialized selection. The shape of the mouth is often a good indication of which strategy a dinosaur primarily adopted. Ankylosaurs like *Euoplocephalus*, for example, had the very wide, flat mouths of generalized eaters and presumably did not discriminate much between one branch and the next after they got stuck into a bush. Small ornithopods like *Hypsilophodon*, with its very narrow mouth, would have been able to choose

CHEWING ACTION. This face-on view of *Edmontosaurus* shows its chewing cycle.

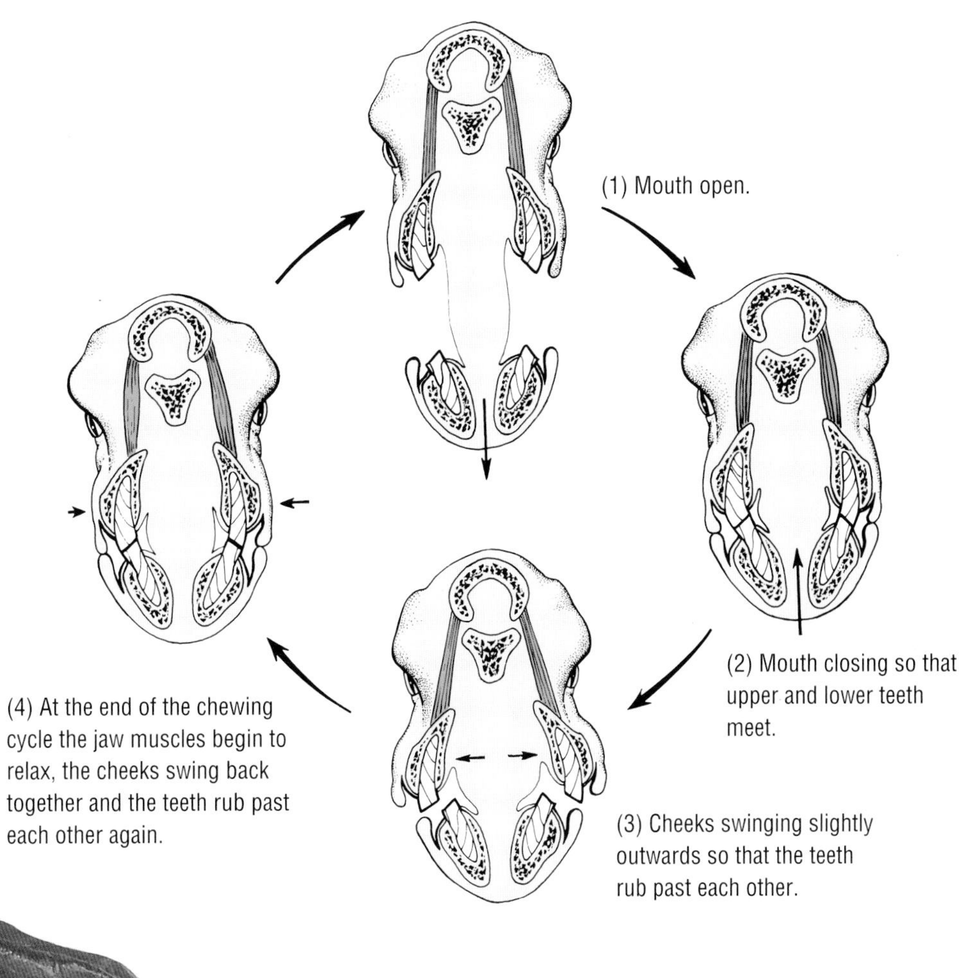

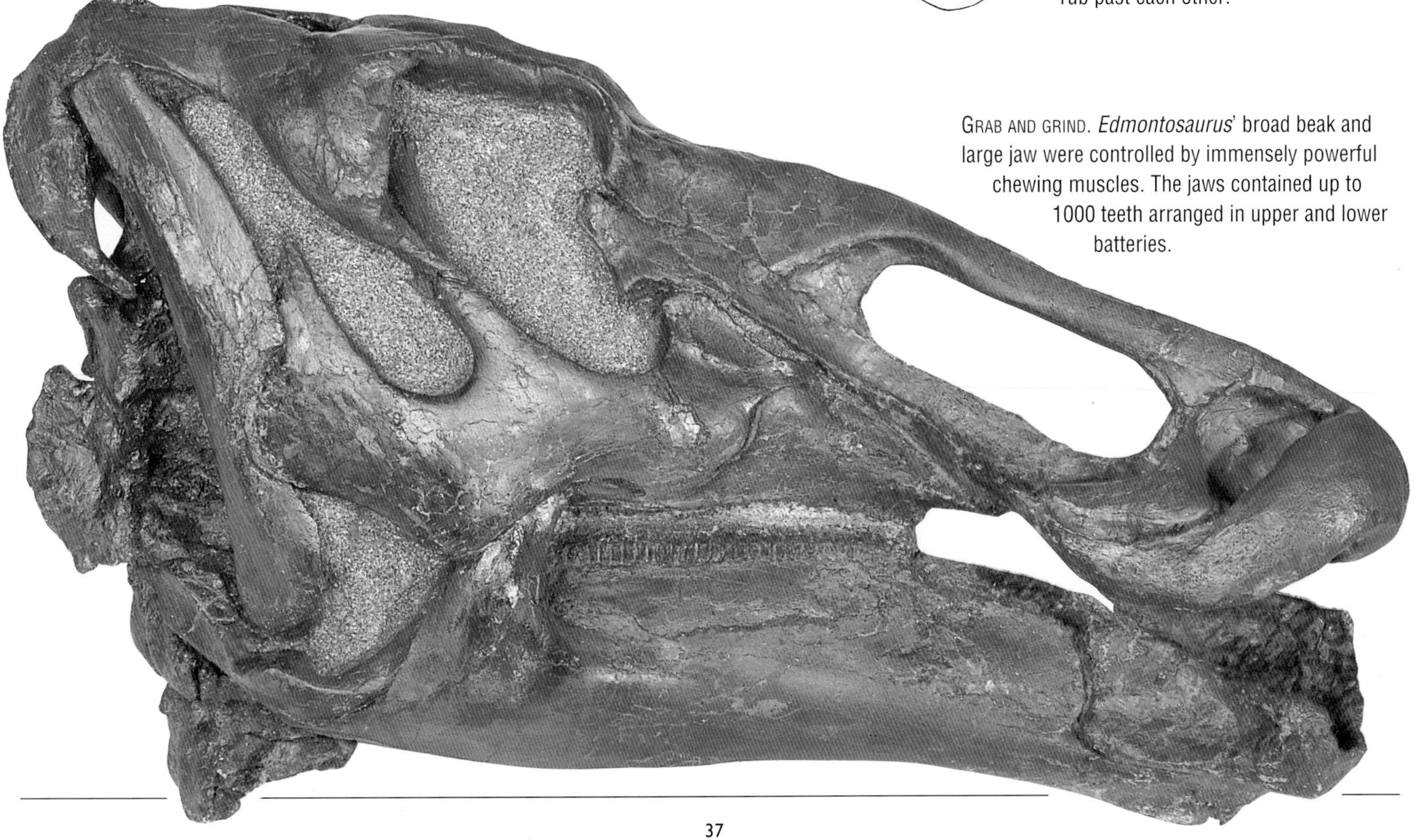

GRAB AND GRIND. *Edmontosaurus*' broad beak and large jaw were controlled by immensely powerful chewing muscles. The jaws contained up to 1000 teeth arranged in upper and lower batteries.

exactly which leaves to accept and which to reject. Modern animals operate in just the same way. The African buffalo's wide jaw is designed to do a quite different job from the narrow one of the antelope, even though they feed on exactly the same areas of land.

The shape of some dinosaur mouths is so specialized that we still do not know exactly what they were used for. The jaws of *Oviraptor*, for example, were probably covered by a horny sheath like a bird's beak. The mouth had no teeth at all but there were two bony tooth-like projections in the roof of its mouth. The lower jaw was also curved to give an extremely powerful bite. Did it use these to crush and feast on the eggs of other dinosaurs? The first specimen ever discovered was just above a nest of *Protoceratops* eggs, giving the dinosaur its name, which means 'egg thief'. Some modern animals, such as snakes, weasels and birds, do steal eggs but few survive on a diet of eggs alone — there simply are not enough to be found. If this was indeed how *Oviraptor* fed, it suggests that lots of dinosaur nests lay unprotected all the year round. Other suggestions are that it dug up molluscs in shallow water and crunched their shells between the powerful jaws or that it cracked open hard fruits or seeds. Whatever the true reason, *Oviraptor*'s unique mouth must have given it access to some kind of food that no other dinosaur could eat.

DEADLY TEETH AND JAWS

The life of a meat-eating animal is very different from that of a plant-eater. When a deer wakes up in the morning it can be pretty certain that the grass will still be there, it is just a case of standing up and starting to nibble. A plant-eater's life, and particularly the life of a plant-eating dinosaur, was really a continuous process of eating with short breaks to sleep, breed, squabble and move on to the next patch of greenery. The meat-eater has no regular food supply, and any meal it does catch may have to last for several days. Most modern meat-eaters, especially big hunters, actually spend most of their time resting (twenty hours sleep out of twenty-four for male lions) with occasional huge bursts of energy for hunting. They gorge themselves on a kill, taking in up to 25 per cent of their own body weight at one feed, and then rest until hunger drives them out to hunt once more. The big meat-eating dinosaurs must surely have followed a similar pattern.

CLIP AND CHOP. *Protoceratops* could use its beak for selective feeding, just as a parrot does. Self-sharpening teeth at the back of the mouth then cut up the plants with a scissor-like action.

NARROW MOUTHS. Narrow mouths allow animals to select food with more care like this kudu. *Hypsilophodon* would have been able to pick out soft shoots and leaves.

BROAD MOUTHS. A wide mouth usually shows a generalized feeder like this cape buffalo, with the emphasis on quantity rather than quality. Ankylosaurs' broad, shovel-like mouths were designed to work in this way.

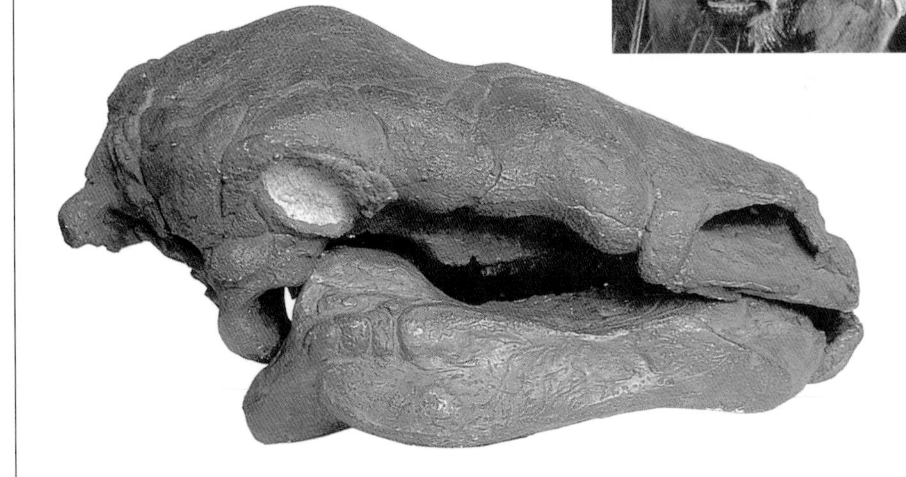

When a big meat-eater like *Allosaurus* went hunting it would head for the herding grounds of *Diplodocus*, *Apatosaurus* and *Camptosaurus*. With luck, it might find the remains of a carcass on the way or a group of smaller meat-eaters gathered round a fresh kill. Few hunters will waste energy on pursuit and killing if some other animal has done the work for them, and all the large meat-eaters were probably occasional scavengers, eating carrion or stealing from smaller dinosaurs as the opportunity arose.

Indeed, some scientists believe that the really huge tyrannosaurs from the end of the dinosaur era were not actually hunters at all but super-scavengers. Among today's animals the only pure scavengers of any size, who do no hunting at all, are vultures and condors. They achieve this by being able to survey huge areas of land from the air, spotting newly dead and dying animals dozens of kilometres away. Obviously the tyrannosaurs had no such ability, so if they were to survive on carrion alone it suggests that there must have been a very large number of dead dinosaurs to satisfy their monster appetites. Bite marks on bones of both *Triceratops* and *Edmontosaurus* leave no doubt as to what *Tyrannosaurus* ate, whether it actively hunted or scavenged. They make it possible to determine that *Tyrannosaurus* was able to exert an extremely powerful, bone-crushing bite, the most powerful known for any back-boned animal. It could generate over 3000 pounds of bite force at a single back tooth. Multiply that by a whole mouthful and the jaws could exert a bite eight times more powerful than a lion. How could the skull withstand such stresses? Emily Rayfield has pioneered techniques used by engineers to find out. She has applied a structural analysis technique, called 'finite element analysis', to measure the stresses and strains on the skull. The skull had quite loosely connected bones around the face, the ligaments that connected them acted as shock absorbers and the very robust bones along the top of the snout provided extra strength. The skull was well adapted to cope with biting and tearing movements and that was exactly how *Tyrannosaurus* tackled its prey — a huge crunching bite followed by tearing backwards through

EGG HEAD. *Right: Oviraptor*'s jaw was designed to exert great pressure, just as the modern parrot uses its curved beak to crush tropical nuts. Did *Oviraptor* use the spikes in the roof of its mouth to break open dinosaur eggs and feast on their energy-rich contents, or did it too feed on hard fruit or perhaps even shellfish?

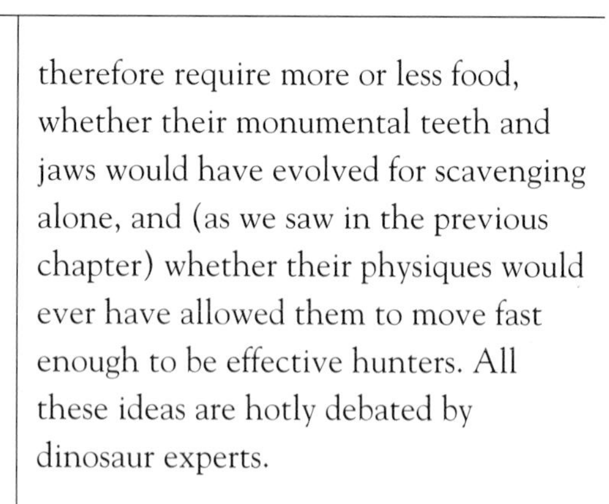

NEW FOR OLD. *Below:* Teeth were replaced throughout the animal's life as they wore down or broke. This *Megalosaurus* jaw shows new teeth growing up on the inside of the old ones.

bone and flesh — a technique called puncture-pull feeding.

One more clue to *Tyrannosaurus*' eating habits was the discovery of an enormous coprolite from 65 million-year-old rocks in Canada. Coprolites are fossilized droppings that can contain the remains of the food its maker has eaten. Various kinds of plant debris are found in coprolites presumed to have been produced by herbivorous dinosaurs. This particular coprolite, the largest know at more than 44 cm long and with a volume of 2.41 litres, contained fragments of bone and must have been produced by *Tyrannosaurus*. The bone fragments had been partly digested by the stomach acids and the bone fibres and blood vessels remaining indicated that a sub-adult prey had been eaten, although it is not possible to identify the victim.

Tyrannosaur's feeding habits raise some of the most basic issues about the large meat-eating dinosaurs — whether they were warm- or cold-blooded and therefore require more or less food, whether their monumental teeth and jaws would have evolved for scavenging alone, and (as we saw in the previous chapter) whether their physiques would ever have allowed them to move fast enough to be effective hunters. All these ideas are hotly debated by dinosaur experts.

Allosaurus is definitely a hunter, and a very hungry one by the time it spots a herd of *Diplodocus* browsing on the tree tops (see pages 32–33). It is looking for

STRONG AND FLEXIBLE. The bones around the sides of a *T.Rex* skull were quite loosely connected to act as shock absorbers. Those along the top of the snout were thick and robust to provide extra strength to cope with the immensely powerful bite.

a vulnerable member of the herd to attack — old, young, sick or disabled are all suitable. *Allosaurus*, like lions or leopards, must have been skilled at spotting the signs of weakness or the subtle changes in the herd pattern that showed one member beginning to stray. Most of hunting is watching and waiting, and *Allosaurus* might shadow the herd for several hours. Hunting alone, it would be essential to stay out of sight, for once the adult *Diplodocus* spotted danger they would instantly gather into a defensive formation, easily able to see off a single predator and possibly inflict fatal damage with their whiplash tails.

Suddenly the moment arrives. One young *Diplodocus* has lagged behind the herd and is now far enough away to risk an attack. *Allosaurus* breaks cover and moves forward, putting itself between the juvenile and the herd. At the sight of the hunter the young *Diplodocus* makes a dash to get back to its parents. Being small, it might be able to outrun the *Allosaurus*, but the hunter is already roaring and edging closer, causing the *Diplodocus* to panic, turn and run. The chase that follows is not long or fast. Neither of these dinosaurs was built for speed and the combined effects of fear and fatigue would soon cause the *Diplodocus* to slow to a terrified standstill.

The scientific term for what happens next is macro-predation. The *Allosaurus* opens its mouth to the furthest extent and runs headlong at its victim. The dagger-like teeth, driven onwards by two tonnes of dinosaur, plunge into the young *Diplodocus*. Its struggles to escape only help the teeth slice further in with their sharp serrations on the front and back edges. *Allosaurus* holds the victim with its front claws and begins to close its jaw, tugging in a series of rapid jerks

SLASH AND TEAR. *Allosaurus* had about sixty curved, dagger-like teeth to kill and dismember its prey. Serrations on the front and back edges made for effective meat-slicing action, while hinged jaws allowed large pieces to be swallowed whole.

GOOD ENGINEERING DESIGN. The green, yellow and red colours on this digital reconstruction of *Allosaurus* show the strengthened areas of the skull that become stressed during biting. Meat-eaters' skulls evolved and adapted on good engineering principles!

with its powerful neck. As a chunk of flesh comes away, *Allosaurus* raises its head and swallows the piece whole. The *Diplodocus* may still be alive but the first priority is to eat the meat as fast as possible. The gorging goes on, with each mouthful torn from the carcass and swallowed whole without any chewing. The front claws, which were much more powerful than those of the later *Tyrannosaurus*, could be used to tear pieces off the carcass. Finally, having consumed a huge quantity of meat, the *Allosaurus* moves off to find a safe spot to rest. There will be no need to hunt again for several days.

Looking more dispassionately at *Allosaurus*, Rayfield's engineering analysis of the skull demonstrated that its structural design was rather different to that of *Tyrannosaurus*. The skull of *Allosaurus* is narrower and more lightly built with strengthening struts at the back. Its teeth are compressed from side-to-side and blade-like, similar to a steak knife, and not at all suited to a crushing bite. *Allosaurus* used its powerful neck muscles to drive a high-speed impact of the skull into its prey, slamming down with its upper teeth to slice deeply through the flesh of the *Diplodocus* and then tearing it off with violent headshakes. Aptly enough, the technical term for this is slash and tear feeding.

Hunters like *Allosaurus* and *Tyrannosaurus* are comparatively rare in the fossil record. Smaller meat-eaters range from the fearsome *Deinonychus* (whose possible pack hunting habits are described in Chapter 5), to the speedy *Compsognathus* which stood no higher than a hen. The smallest meat-eaters must have hunted lizards, mammals and large insects. For smaller meat-eaters, insects represent highly concentrated packages of nutritional energy. Indeed, the juveniles of many modern reptiles start on a diet of insects until they are

DINOSAUR DIETS. Some fossil skeletons give evidence of what the dinosaur last ate. This *Coelophysis* has the bones of its own young in its stomach, the only known example of dinosaur cannibalism. The tiny vertebrae and one thigh bone can be seen between the ribs of the adult.

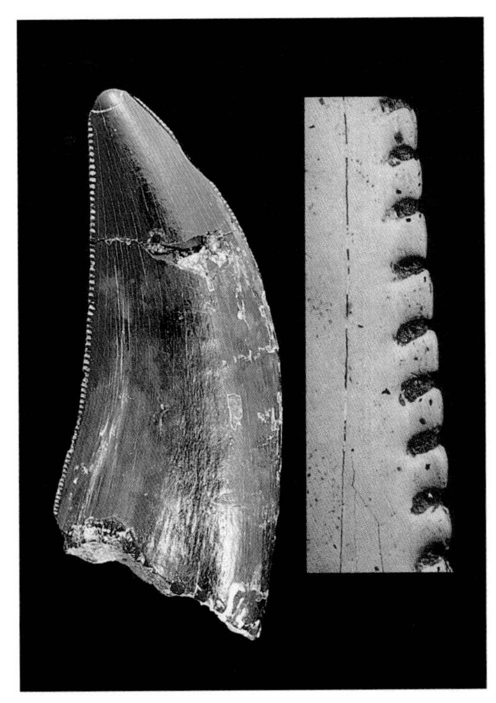

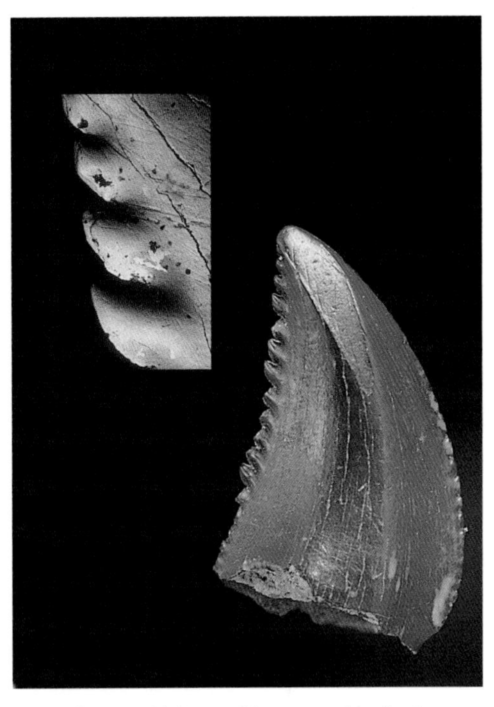

TOOTH TOOLS. Small, even denticles provided a sharp cutting edge on allosaur teeth (*left*); the edges of tyrannosaur teeth were thicker and bone-crushingly strong (*centre*) and the coarse, spiky edges of troodont teeth were suited to slashing or shredding flesh, or possibly plant material (*right*).

big and fast enough to capture other prey, so perhaps juvenile dinosaurs did the same. Meat-eaters like *Velociraptor* and *Dromeosaurus* were clearly fast runners with bodies characterized by light bone structure and long legs, while powerful claws and rows of needle-sharp teeth helped them get a firm grip on the squirming victim.

It is not possible to deduce from the evidence currently available what sort of hunting each of these many meat-eaters specialized in. Particular refinements of their anatomies, especially teeth, do give clues as to how prey was processed, however it was hunted. Meat-eaters' teeth are generally recurved and blade-like with denticles along the front and back edges like the serrations on a steak knife. There are many variations on the basic pattern. *Allosaurus* and *Carcharodontosaurus* had compressed, quite flattened teeth with even-sized denticles. They were efficient meat slicers, the denticles forming a sharp cutting edge. The properties of tooth enamel and dentine are such that teeth cannot be as thin and sharp as a metal knife blade and at the same time strong enough to resist the stresses of a struggling prey. Serrated edges apparently allow a tooth to cut as efficiently as a thinner smooth blade would do, while being stronger and less likely to break. *Tyrannosaurus* teeth were much thicker and stronger, quite coarsely serrated and more suited to gripping, tearing and crushing functions. *Troodon* had relatively small, strongly recurved teeth with large jagged denticles on the back edge – efficient rippers as the teeth were pulled backwards through the victim's flesh. Occasionally a fossil skeleton is found with the remains of the dinosaur's last meal still inside. Thus we know, for example, that *Coelophysis* was a cannibal, since the bones of several babies have been found inside the stomach of an adult. *Compsognathus* from Germany was discovered with the half-digested remains of a speedy lizard called *Bavarisaurus*, while the recently discovered British dinosaur *Baryonyx* was found to contain scales and teeth of the large, non-marine fish *Lepidotes*, making it the first known fish-eating dinosaur. The story of *Baryonyx* and its relatives is told in Chapter 9.

The ability to obtain and process food is key to any animal's survival and a major factor in determining its evolutionary development. The dinosaurs were overwhelmingly dominant on land and grew in some cases to an enormous size. These facts alone are a tribute to how efficiently they managed to exploit the available food supply and turn virtually every land habitat into a potential feeding ground.

ATTACK AN

Just like the animals of today, dinosaurs attacked each other, defended themselves and competed for leadership within groups. The equipment they used was sometimes

D DEFENCE 4

spectacular and lethal, yet for every giant hunter like *Tyrannosaurus* there were many more small plant-eating dinosaurs which did nothing more aggressive than running away.

The largest hunter ever to walk the Earth was *Tyrannosaurus*. A pack of *Velociraptor* could have ripped a human being to pieces in 30 seconds and *Allosaurus* swallowed you in two mouthfuls. Yet the dinosaur world was not actually dominated by roaming gangs of monster killers. Most dinosaurs were peaceful, plant-eating creatures that never attacked anything in their lives. The weapons they carried were either used to defend themselves or to compete for food, space and a mate within their own group. And while it is the attackers that usually make headlines in dinosaur books, the defenders and competitors were just as spectacular in their own right.

Modern animals have many ways to defend themselves — disguise, running away, herding together, fighting back. Strategies range from the active use of teeth, claws and horns to passive weapons like camouflage and armour. One dinosaur with a remarkable combination of active and passive defences was *Euoplocephalus*.

CLUBS AND ARMOUR

Euoplocephalus was a four-legged plant-eater, cropping low growing vegetation of the North American ranges with its toothless beak. Every part of its back, neck and tail was armoured with bony plates set into the skin. These varied from huge blunt spikes and slabs across the neck and shoulders, down to thousands of pebble-sized lumps like chain-mail along the tail. The head was even more heavily armoured, with slabs of reinforcing bone over all exposed surfaces and triangular side studs above and below the eyes. Even the eyelids were made of bone, closing like steel shutters to deflect the gouging claw of an attacker.

WEAPONS AND THEIR USES. Attack, defence and competition are keys to any animal's survival. *Top:* Hunters combine speed and strength with weapons that kill quickly. *Middle:* Tough armour provides excellent passive defence for an armadillo. *Bottom:* Bighorn sheep butt heads in the mating season. This is more bluff and display than real fighting.

Euoplocephalus' passive defence strategy was simple. Crouching down to protect its soft belly, it presented a completely armoured shell, with every part of the body protected. No amount of slashing or biting could make any real impression, and the only option for an attacker was to try and flip *Euoplocephalus* over on its back — about as hard as overturning a 2 tonne slab of rock, particularly when its front legs were 'locked' into a defensive position against its shoulders.

Despite the huge weight of all this armour, perhaps half a tonne in all, *Euoplocephalus* seems to have been surprisingly agile. Muscle scars round the hips and shoulders suggest it had enormously strong legs which gave mobility as well as controlling the movement of a giant tail club. For, while armour provided passive protection, *Euoplocephalus* also had a devastating active defence weapon that could transform it from a frustratingly

(*Previous spread*) A KILLER EYE. Any dinosaur that got this close to a *Tyrannosaurus rex* probably never saw anything else again.

inaccessible meal into a highly dangerous adversary.

The tail of a fully grown *Euoplocephalus* was around 2.5 metres long and ended in a huge club made of two fused bones weighing up to 30 kilogrammes. By lashing out from side to side, the club could deliver a scything blow to the legs, easily enough to smash bones or send the largest tyrannosaur crashing to the ground. Once down with a broken shin or worse, the attacker never rose again and soon fell victim to other hunters and scavengers on the prowl. In its active defence stance we can imagine *Euoplocephalus* being circled by a wary tyrannosaur, looking for an opening for the quick hind leg kick that might tip the smaller dinosaur over or disable it. All the time *Euoplocephalus* circles as well, turning its back towards the attacker with tail poised to strike. It would be a very confident or a very hungry hunter indeed that felt like taking on such an opponent.

OFFENSIVE WEAPON. A tyrannosaur circles a *Euoplocephalus*. One blow from the tail club could be fatal.

DEFENCE AND COMPETITION

For large sauropod dinosaurs like *Brachiosaurus*, size alone was usually sufficient defence. As with elephants today, there were no predators large enough to tackle a fully grown adult. All sauropods also had a pair of thumb spikes. In the case of the rather smaller *Diplodocus* these could perhaps have been brandished while it reared up backwards, leaning on its tail. The tail itself was also a formidable whiplash weapon — equivalent to 13 metres of flailing telegraph pole.

At the other end of the scale, the skeletons of small plant-eaters such as *Dryosaurus* show that running away was often the best form of defence. Good hearing and eyesight must also have been essential for rapid flight at the first sign of danger.

BUILT FOR DEFENCE: *Euoplocephalus*.

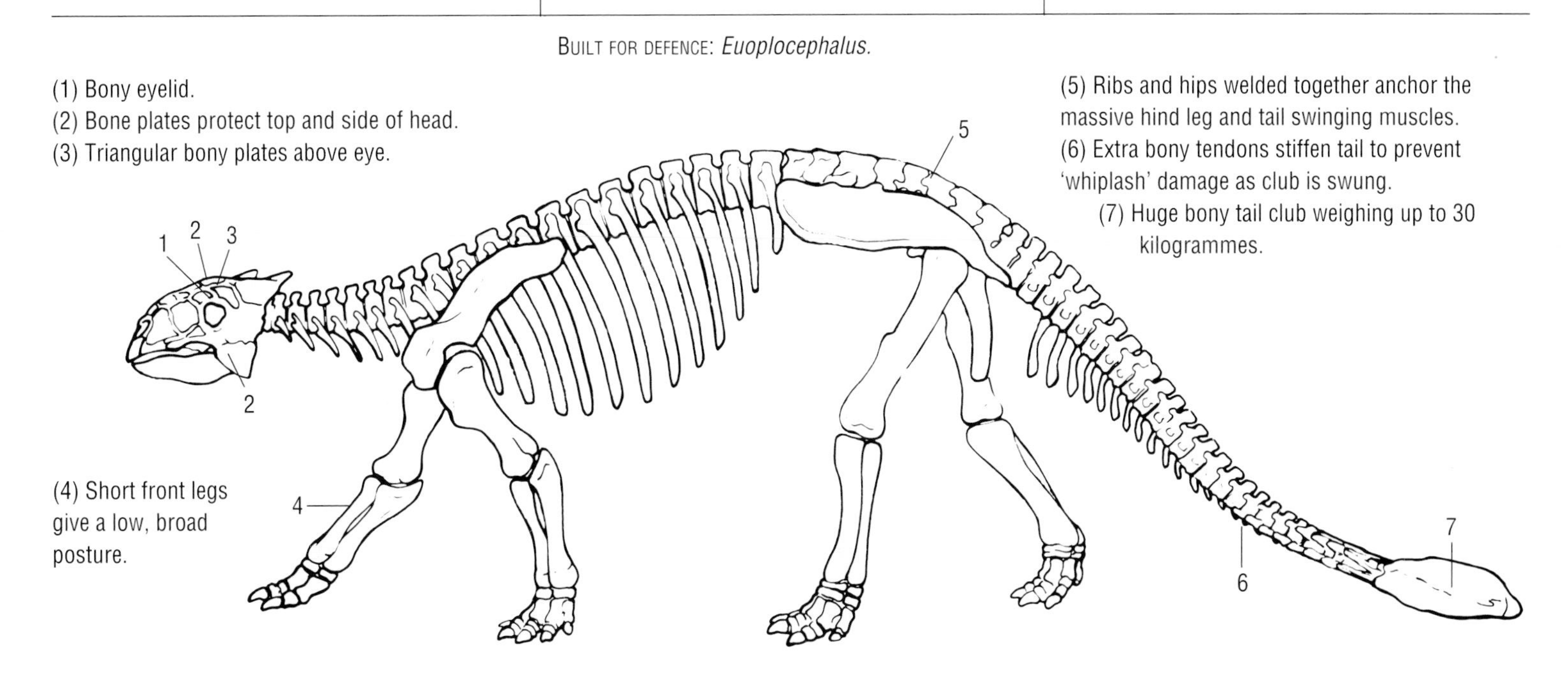

Diplodocus' claw and tail, *Dryosaurus*' speed and the armour on *Euoplocephalus* are highly developed versions of defence strategies found extensively throughout the dinosaur world. Plant-eaters would usually have tried to escape before fighting. Many protected vulnerable parts of their bodies, such as necks and spines, with bony neck frills, tooth-snapping bone studs and plates, and many had a sharp claw, horn or spike for swift counter-attack. For dinosaurs, like today's animals, defence did not usually mean being totally immune from any attack but slightly harder to catch or to kill than other prey nearby. A balance of active and passive defences to keep one step ahead, sometimes literally, was the way that most dinosaurs developed.

Dinosaurs that lived in herds for protection would almost certainly have competed with each other. Many gregarious and social animals operate within a firm social structure. Whether it is apes or insects, there must be a clear mechanism to fix the function of every member (leaders, followers and partners) if the group is to survive. Sometimes this mechanism is genetically programmed in the animal, but in other cases social order is determined by competition between group members. The strongest elephant leads the herd, the smartest looking peacock gets a mate. Such competition, even if it involves actual fighting, very seldom results in death or serious injury. The purpose after all, is to establish order in the group, not reduce its overall strength. Many herding dinosaurs had features, such as crests or bony frills, that suggest social hierarchy. In one case, however, the competitive feature is so remarkably developed that the dinosaur is actually named after it, *Pachycephalosaurus*, the 'thick headed lizard'.

Pachycephalosaurus was a two-legged plant-eater that lived in western North America during the late Cretaceous Period. At first glance it looks like a dinosaur genius. Surely that high domed head housed a mighty brain! But no, the head was solid bone, up to 25 centimetres thick, providing protection for a brain no bigger than a tennis ball. Protection from what? From the crunching impact of a head butting contest with another *Pachycephalosaurus* as they did battle to establish social order within the herd.

The evidence for this activity lies chiefly in the structure of the skull and spine of *Pachycephalosaurus*' smaller relatives from North America and Asia. The off-centre way that the head met the neck allowed the animal to adopt a head down charging position when walking or running. Strong ligaments joined the skull to the back of the neck, absorbing the shock and transferring it to the spine. The spine itself was strengthened with bony rods and had a special grooved joint between each vertebra to prevent twisting under pressure. There may also have been air spaces around the brain to reduce the jolt it received. All this would have enabled two male *Pachycephalosaurus*, each as long as a bus, to hurtle together and clash heads at truly terrifying speed without doing fatal damage.

Modern techniques have cast doubt on such a scenario. Analysis of thin sections of dome bone from individuals of different ages suggest that the bone was spongy in young animals, but this disappeared in adults so there would be no shock-absorbing effect. This has prompted Mark Godwin and John Horner to propose that the domes functioned for species recognition and communication in adults, and possibly for sexual display, but this does not

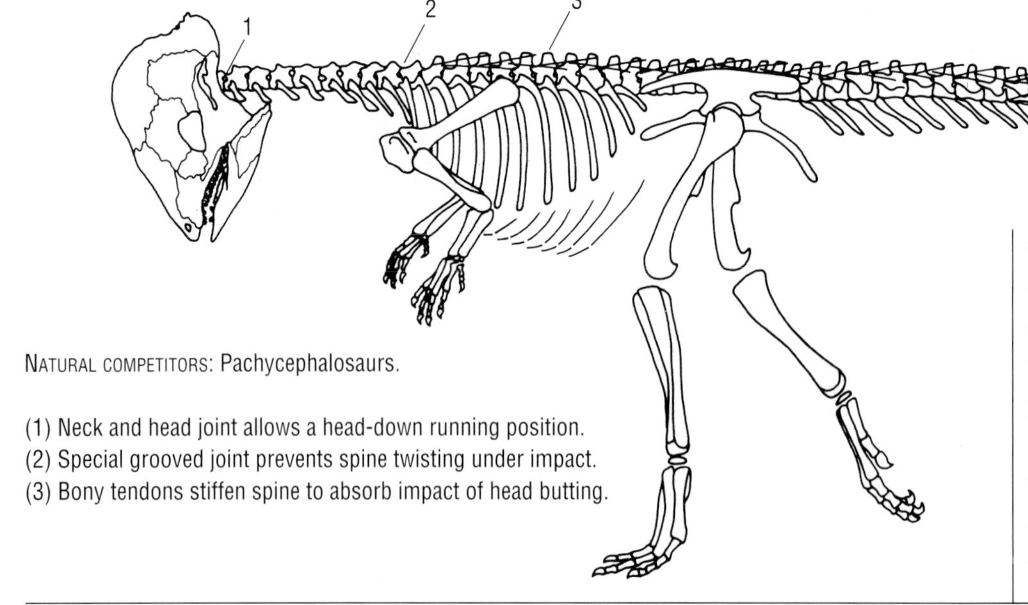

NATURAL COMPETITORS: Pachycephalosaurs.

(1) Neck and head joint allows a head-down running position.
(2) Special grooved joint prevents spine twisting under impact.
(3) Bony tendons stiffen spine to absorb impact of head butting.

KILLED IN ACTION. This *Protoceratops* and *Velociraptor* from Mongolia are the only example of dinosaurs discovered in combat. The *Velociraptor*, its chest crushed, still had its claws gripped behind the enemy's neck frill. Just below the bodies was a nest of *Protoceratops*' eggs, so perhaps the *Velociraptor* was caught in the act of robbery. Both dinosaurs were overcome by a sandstorm as they fought, and died still locked together.

explain the unusual features in the neck joint and vertebrae.

Probably all herding dinosaurs used noisy bluff, display and trials of strength in one way or another to organize themselves. For it was only in a coherent, well-ordered group, staying close to a leader, looking after their own family members and swiftly following direction, that any herbivores could hope to survive the attentions of the many attackers that lay in wait for them.

A STREAMLINED KILLER

If any dinosaur deserves to be described as built for attack it is *Deinonychus*, a streamlined killer that combined agility and speed with ferocious weaponry. Standing about 2 metres tall and up to 3 metres long, *Deinonychus* had the light, long-legged physique of a fast runner. A sheath of bony rods along the tail made it stand out stiffly behind. This acted as a counterbalance when sprinting after a victim, giving extra manoeuvrability. *Deinonychus* had a slim neck and lightweight skull, ideal for quick, snapping bites. Teeth were backwardly curved for efficient meat slicing action and used in combination with the front and hind claws. The arms acted as long grappling hooks, grabbing a small creature to keep it still or allowing *Deinonychus* to leap up and hang on to a larger dinosaur. The second claw on the hind foot was like a crescent-shaped slashing knife. It has been suggested that a blow, delivered with a massive, sweeping kick, would have been enough to rip open the belly of a victim or tear it in half. However, tests led by Phil Manning using a robotic model of the hind limb and a pig carcass, show it produced small puncture holes, but no ripping or slashing marks. A comparison of the *Deinonychus* claw with a range of modern claws from reptiles, birds and mammals, showed that neither the claw bone or the keratinous sheath that covered it had a sharp cutting edge on the under surface. The extreme curvature of the claw proved to be similar to those of tree trunk climbing birds, which use them for clinging and gripping.

It seems that *Deinonychus* used these claws as climbing aids — like a climber's crampons — to hook onto and clamber up the skin of their prey, in combination with their grasping hand claws, while their sharp, serrated teeth did their deadly work. The whole foot was adapted to keep the big claw in good condition. When walking or running it was pulled up and back to avoid blunting on the ground. When attacking, it could swivel through 180° to find the best possible angle of entry.

When the first *Deinonychus* fossils were found in Montana in 1964, scientists noticed that there were the remains of several of the predators near the body of a much larger herbivore, *Tenontosaurus*. Did they hunt in groups? Certainly *Deinonychus*' brain was relatively large and probably well developed in the areas vital to a pack hunter, namely those of sight and sound. Bringing down a big dinosaur would have provided several days' food supply, even if it had to be shared.

The dinosaur era produced many fine hunters, but in *Deinonychus* we see the near perfect balance of an attacker's skills. Intelligence to stalk, ambush or mount a co-ordinated attack, the speed to run a quarry down and the weapons to kill and dismember it.

Bony rods stiffen tail to counterbalance while sprinting.

Many of the other meat-eating dinosaurs developed one of these qualities at the expense of the others. *Tyrannosaurus*, for example, depended on bulk and enormous strength to overpower its prey. Its head was heavily reinforced with bone and shock absorbing muscle to withstand the impact of crashing into a victim with mouth open and then delivering a crushing bite and neck twist. But with a body weight of 7 tonnes *Tyrannosaurus* had clearly sacrificed speed to size. If it did not catch its quarry in the first rush there was no question of setting off in a lengthy pursuit.

At the other end of the scale, *Troodon* developed speed and brain power at the expense of size and weaponry. A small, fleet-footed predator, *Troodon* had a large brain and front facing, stereoscopic eyes, probably allowing it to stalk rat-sized mammals that came out at dusk, when other dinosaurs could not see to hunt.

Euoplocephalus, *Pachycephalosaurus* and *Deinonychus* all bear the evidence of their way of life clearly written in their bones. Yet the fossil record tells us little about the attack and defence behaviour

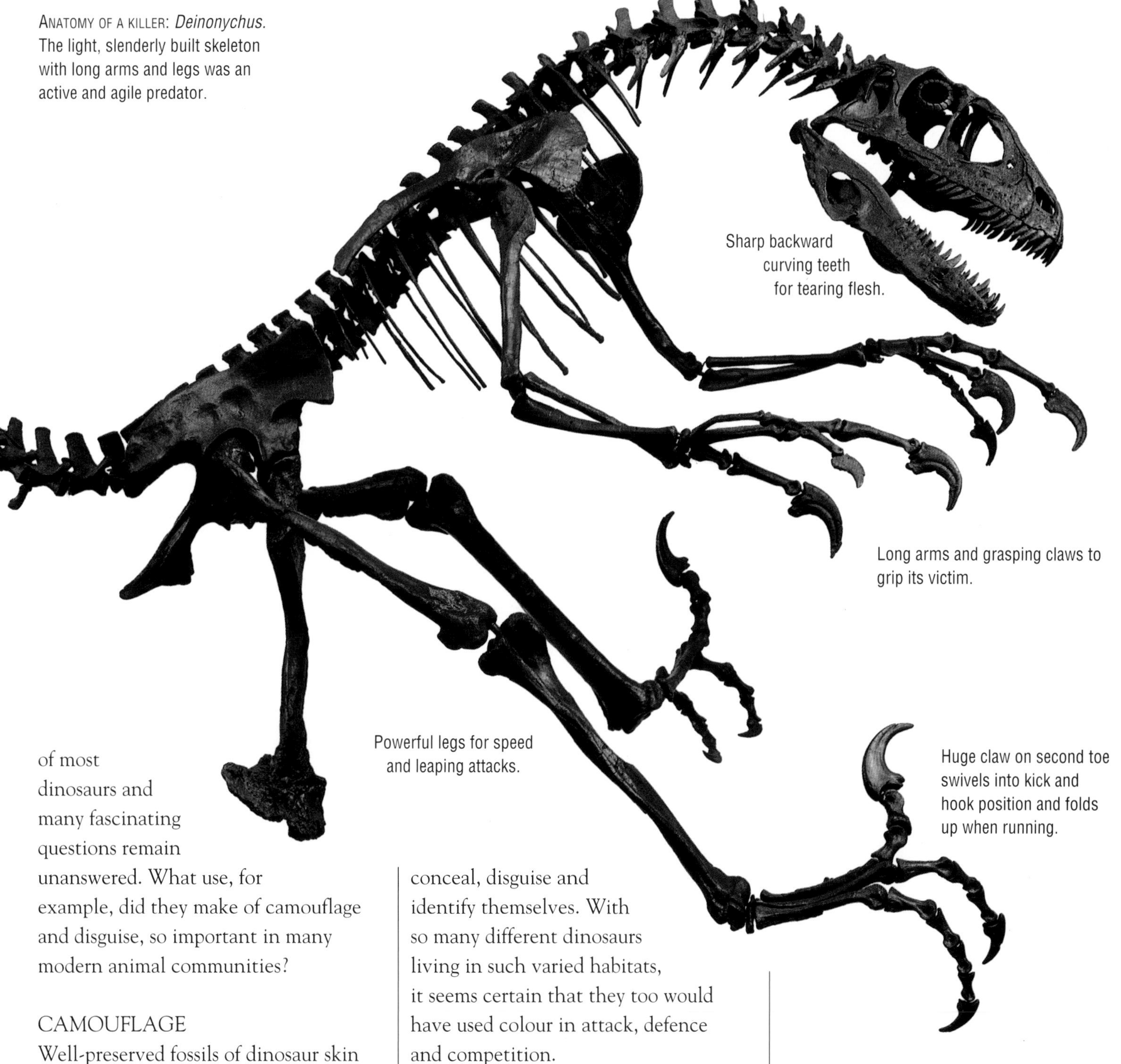

ANATOMY OF A KILLER: *Deinonychus*. The light, slenderly built skeleton with long arms and legs was an active and agile predator.

of most dinosaurs and many fascinating questions remain unanswered. What use, for example, did they make of camouflage and disguise, so important in many modern animal communities?

CAMOUFLAGE

Well-preserved fossils of dinosaur skin are very rare but in some cases sections of the skin did become dried out and preserved, perhaps because the animal was rapidly buried by a sandstorm that hid it from scavengers and mummified the body. These scraps of evidence show that dinosaurs were covered with tough, scaly skin like modern reptiles. We know that colour helps modern animals to conceal, disguise and identify themselves. With so many different dinosaurs living in such varied habitats, it seems certain that they too would have used colour in attack, defence and competition.

A medium-sized hadrosaur like *Kritosaurus* would have spent much of its time browsing on ferns and low bushes under the forest canopy. A dappled skin colour of greens and yellows, like an anaconda, would have been ideal to camouflage its low body among the sunsplashed vegetation on the ground.

When it sensed danger *Kritosaurus* might well have frozen still, just as bushbuck do, in order to blend into the forest and remain unnoticed. *Parksosaurus*, browsing in the same forest, might perhaps have developed a more definite skin pattern like a striped monitor lizard to keep itself concealed as it passed through broken areas of light and shade.

VARIED DISGUISE. An anaconda, okapi and monitor lizard show three modern approaches to the use of colour in attack and defence. Do they give a clue to the way that dinosaurs like *Kritosaurus*, *Troodon* and *Parksosaurus* may have looked?

Bold countershading is the disguise used by the mainly nocturnal okapi browsing in areas of dense undergrowth. The dark back, and its striped and light rump and upper/lower legs, serve to break up the body shape when seen from a distance, providing protection against predators like leopards that hunt by sight. Perhaps the fast moving hunter *Troodon* had similar colouring in order to conceal itself.

Colour must also have played a key part in dinosaurs' competitive behaviour. Many male animals today, particularly small birds and reptiles, adopt bright identifying colours during the mating season. The colouring announces that the male is ready to breed, helps females make their choice of partner and can often be linked to clashes between competing males.

Perhaps crested hadrosaurs such as *Corythosaurus* and *Lambeosaurus* took on special head colouring as their herd approached its mating ground. *Saurolophus* and *Edmontosaurus* may have used their fleshy nose sacks rather as an elephant seal does to produce booming mating cries. Possibly the sack itself became brightly coloured. The head structure of *Styracosaurus* also hints at display and competition. Its round bony frill was actually longer than the skull itself, and decorated round the outer edge with long spikes. Yet the frill was not solid bone like *Triceratops*', but contained two large openings covered with muscle and skin. By tilting its head forward and swinging it from side to side, *Styracosaurus* could produce an impressive display of size and aggression to frighten off other males and advertise its prowess. Two male *Styracosaurus* with horns locked like stags' antlers could engage in pushing, barging and bellowing contests that would produce spectacular and noisy proof of virility without any real injury being done. Once again, the skin colouring of the frill might have changed to a distinctive pattern at mating time.

Beyond the painstaking deduction based on firm fossil evidence, it is fascinating to speculate about the true behaviour of dinosaurs in situations of attack, defence and competition and to imagine them as vividly coloured as today's birds and reptiles, their nearest living relatives.

HORNS AND SPIKES. *Styracosaurus* seen head-on would present a formidable sight to an attacker.

SOCIAL ORG

Social organization remains one of the great mysteries of dinosaur science. Yet, as living animals, they must have displayed patterns of behaviour similar to those we see in

ANIZATION 5

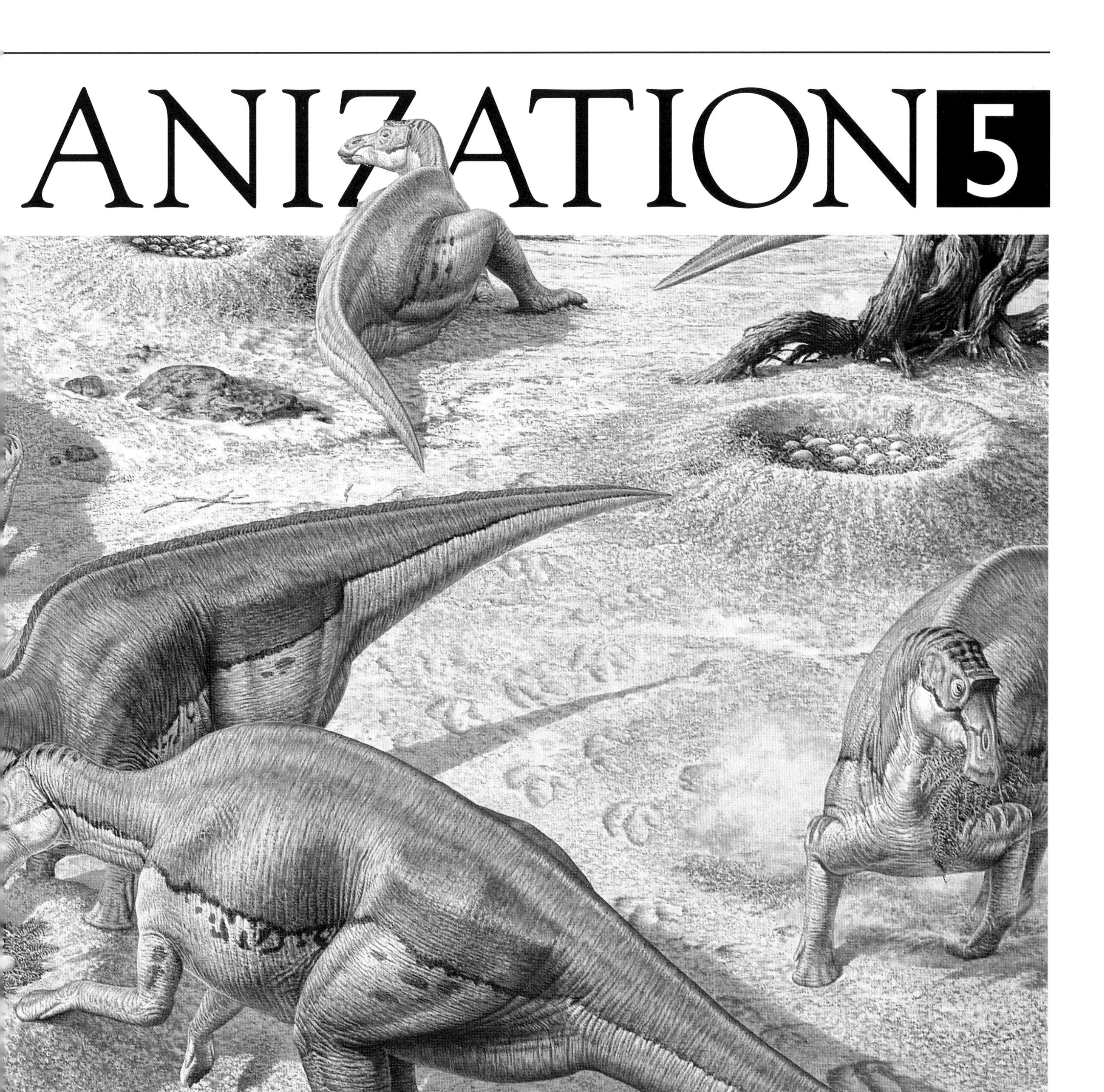

today's animal world. Recent discoveries of nesting colonies have shown that some dinosaurs achieved a high level of social organization.

If any aspect of prehistory requires the use of a time machine, it is the study of dinosaurs' social lives. Biologists can spend many years with communities of lions, gorillas or birds to observe how they live. Only such detailed fieldwork can give a true insight into complex family structures, group behaviour and techniques for survival. You cannot do much field observation on a fossil skeleton, so palaeontologists must draw evidence from a wide range of sources to gain a picture of dinosaurs' social behaviour.

The results are mostly speculative with the words 'perhaps', 'possibly' or 'maybe' tending to appear in virtually every sentence. Some experts prefer to avoid this area of study altogether, yet dinosaurs were social animals operating in herds, packs, families and at the very least mating pairs. They shared the same land — hunting and avoiding each other, competing for space and food, just as modern animals do. Social behaviour is as important to understanding dinosaurs as looking at their bones and teeth under a microscope.

If we think of dinosaurs as members of a widely spread, balanced community, with each member dependent on the others, then the closest parallel in today's world are the African savannas. The foundation of these communities are the plants (grass in the savannas but *not* in the dinosaur world; see Chapter 1), which provide food for numerous plant-eaters. These in turn are food for the hunting meat-eaters which live near the fringes of their herds, catching what they can. It takes a lot of plant-eaters to support a meat-eater. One lion, for example, needs access to at least one hundred gazelle or zebra to provide enough hunting opportunities for survival. Behind the hunters come the hyaenas and jackals, animals that often live off the leftovers of the big kills, and occasionally do some killing of their own, and the true scavengers such as vultures.

There are many levels of interdependence between these different animal groups. It is obvious that lions need zebra and gazelle to eat. It is less obvious, but equally true, that zebra and gazelle need lions to keep their numbers down and prevent huge herds from overgrazing the range and causing mass starvation. Elephants consume a huge amount of food each day, enough to feed scores of smaller oryx and wildebeeste. But oryx and wildebeeste need the elephants, whose amazingly destructive eating habits — uprooting and trampling — open up the forest for bush animals and the denser bush to plains dwellers. Antelopes share their grazing grounds with ostriches because ostriches are such good lookouts, and so on. In the dinosaur world too there must have been many such examples of interdependence between groups that chased, fled, avoided or shared territory with each other.

THE LONE HUNTERS

Except in very rare cases, no skeletons of big hunting meat-eaters have been found near each other. This is in contrast to many 'mass grave' finds of herds of plant-eaters, and suggests that some, at least, of the large hunters operated alone. This is good survival logic, since there is no point in cutting down your food supply by sharing a hunting ground with another large stomach. The carnosaurs would have had little to gain from co-operative hunting as they were so much bigger than many of their potential victims and would have needed no help to kill them. For a single hunter the best places are forests or dense scrub, since this gives plenty of opportunity for ambushing animals which would certainly spot you coming on a flat plain. The modern leopard uses this lone hunting technique.

There are some areas where the number of meat-eating dinosaur trackways, both large and small, is surprisingly high. These may have been favourite hunting spots, perhaps along the shores of rivers and lakes, where the hunters constantly patrolled to pick up carrion that drifted ashore, or attack plant-eaters as they came to drink.

However lonely a life the big hunting dinosaurs may have led, they must have got together to produce offspring. Again, we have very little evidence of their mating habits, although eggs attributed to meat-eaters have recently come to light in the Gobi Desert. Did they nest

---

(*Previous spread*) A NESTING COLONY. A colony of the hadrosaur *Maiasaura* looking after their young in the nest. The dinosaur's name means 'good mother lizard'.

in groups or alone? One clue could be the discovery of different meat-eaters' skeletons with the bones of their own young inside their stomachs. Among today's animals, male bears and lions will occasionally try to kill off the young of other males, and some reptiles will actually eat their own hatchlings if put under stress. If cannibalism was part of the hunting dinosaurs' behaviour they would have tended to nest alone. They might also have been female dominated with the male being kept well away from the nest by the aggressive mother, and even being seen off completely after successful mating. The hatchlings would probably have stayed with the parent, or parents, during their most vulnerable early months, eventually leaving to find their own territory. A tantalising recent fossil find seems to show an adult *Tyrannosaurus* in association with a smaller adult, a juvenile and a baby. This may prove to be the first evidence of a carnosaur family group.

LONE HUNTERS. Hunt alone like a leopard or *Tyrannosaurus* and you do not have to share the kill. A lone hunter must be strong enough to overwhelm its prey (here an unfortunate *Edmontosaurus*) and then keep other scavengers away from the carcass.

THE PACK HUNTERS

There is good reason to believe that some of the smaller meat-eating dinosaurs lived and hunted in packs. Co-operative hunting is practised by many of today's animals with different degrees of sophistication. A surging army of soldier ants will make a concerted attack on a wasps' nest and destroy it with great efficiency. This is certainly co-operative, although the level of interaction it requires is pretty low. At a more complex level, pelicans fish together, and lions have complex techniques of encircling and stalking that may include several individuals operating as a highly organized team. The advantage of co-operative hunting is clear — the ability to get more and bigger prey than can be achieved alone. Some pack hunters can bring down a victim as large as their total combined weight. The disadvantage is having to share the kill, but even this is not necessarily very serious. With a whole pack feeding from a carcass very little goes to waste through rotting, and there is no need to defend the kill over a period of hours or days from the attentions of other scavengers as it is quickly devoured.

DINOSAUR HERDS. *Above:* Multiple trackways are evidence that some plant-eating dinosaurs lived and moved in large herds. This picture shows a herd of *Centrosaurus* in a defensive circle. *Right:* A mass grave of *Centrosaurus* shows where a herd crossed a swollen river and many individuals were swept away.

As mentioned in Chapter 4, the first find of the fossils of the meat-eater *Deinonychus* suggested that they might have been hunting in a pack to bring down a *Tenontosaurus*. How many were there in the group, and what hunting techniques did they use? It is easy to imagine several of them gripping on to *Tenontosaurus*' tail to slow it down while the pack leader climbed up to deliver slashing bites to the flank and stomach. Or did *Deinonychus* hunt in smaller, family-based groups?

Modern African hunting dogs give an indication of how effective such co-operative hunting can be. The degree of communication required to hunt together often extends into the complex social interaction between pack hunters. Before each hunt a highly ritualized series of sounds and movements binds the African hunting dog pack together, reinforces its social order and sends it on its way. They also share the spoils of the hunt with their young pups as well as old and sick members not able to join the chase. Pups old enough to run with the hunt are allowed to eat their fill before the adults finish off the carcass.

The social life of *Deinonychus* makes a tantalizing case study. There is just about enough evidence to place them as pack hunters. It is reasonable to assume that co-operative hunting was reflected in some form of co-operative lifestyle that encompassed mating, rearing young, migration and movement as well as just attacking prey. Yet we are left with many unanswered questions. It is not until we come to the herding plant-eaters from the end of the dinosaur era that we start to find some really convincing proof of the complexities of dinosaurs' social behaviour.

THE HERDING DINOSAURS

There have been enough discoveries of groups of skeletons together to be sure that the plant-eating hadrosaurs and ceratopians of North America lived and moved in large herds. Some of the finds are literally mass graves, with one 'bone bed' in Alberta, Canada, containing the fossilized remains of at least 300

Centrosaurus of all sizes, from yearlings to large adults. The composition of the surrounding rocks suggests that there might have been a deep river in the area and the *Centrosaurus* herd was swept away trying to cross it. An even more extraordinary mass grave in Montana contains the bones of at least 10,000 *Maiasaura* whose herd was overcome by volcanic gases and buried in ash. The bed of preserved bones stretches in a solid line for almost 2 kilometres. Other less spectacular mass finds still show members of the same group or species sharing land area, adults and young together. So it is reasonable to look at these particular dinosaurs for further evidence of family and group behaviour.

Some hadrosaurs had varied head crests which invite comparison with the many different horn shapes of antelope in the African savannas. For these hadrosaurs, as for antelope, the fancy headgear had several uses. Its distinctive shape gave a means of instant recognition to members of the same species — very important if you need to move off with your herd in a hurry. Antelope horns also distinguish between male and female, and there is some evidence that hadrosaur crests did the same. Detailed measurement of the fifty or so crests so far found, shows that there were large and small versions within each species. Some hadrosaurs may also have been able to use their crests as sound making devices. Dr David Weishampel of Johns Hopkins University, Baltimore, USA, has built a full-scale model of the long crest of *Parasaurolophus*, reproducing exactly the size and shape of the internal tubes. By blowing air through as *Parasaurolophus* would have done, the model produces a low note rather like an oboe. Such a sound, and variations on it, would give plenty of scope for communication and aggressive bellowing at mating time. It is also ideal for sounding the alarm to the herd when a predator is spotted. Low frequency sounds offer two important advantages, firstly, they carry well over long distances and secondly, it is very difficult to locate their source. This latter advantage would have been very important to a *Parasaurolophus* which did not wish its warning signal to make it the centre of attention for an approaching attacker.

Besides self-protection, the main purpose of any animal herd is to provide a suitable environment to produce and raise young. Until recently the only real evidence of dinosaur family life were eggs and nests from the Gobi Desert discovered in the 1920s – the first real evidence that dinosaurs hatched from eggs – and some trackways that showed adults and juveniles moving together. Then a series of discoveries by Dr John Horner in Montana, USA, completely transformed our understanding of how certain dinosaurs reared their young, and shed new light on the sophistication of dinosaur social behaviour.

More than two hundred dinosaur egg sites are now known from around the world. Most of them occur in late Cretaceous sediments deposited in arid or semi-arid environments. Some dinosaurs laid their eggs on beaches: a huge concentration of sauropod eggs and shells in the southern Pyrenees provides evidence that favoured areas were heavily used and nests, containing between one and seven eggs, were very

DISTINCTIVE HEADS. The different shapes of hadrosaurs' crests might have allowed them to recognize members of their own species and the sex of individuals.

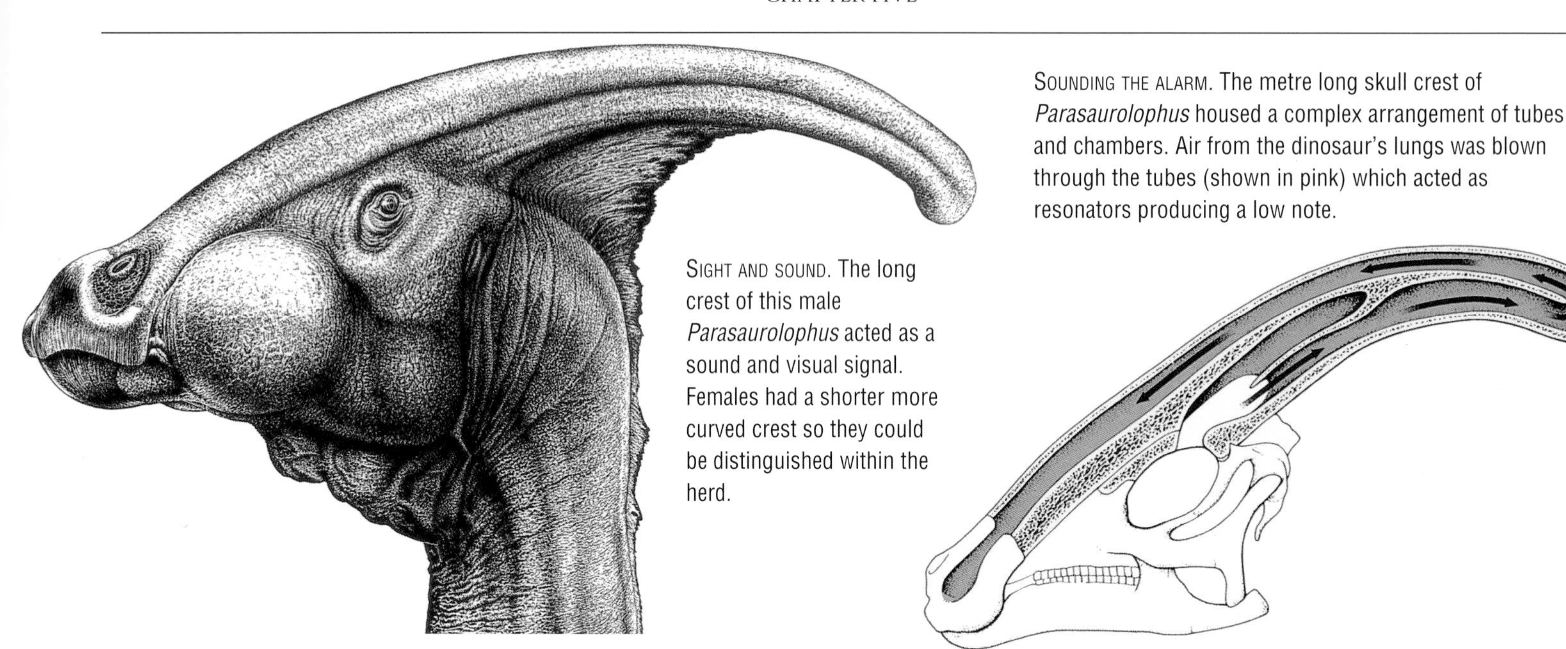

SIGHT AND SOUND. The long crest of this male *Parasaurolophus* acted as a sound and visual signal. Females had a shorter more curved crest so they could be distinguished within the herd.

SOUNDING THE ALARM. The metre long skull crest of *Parasaurolophus* housed a complex arrangement of tubes and chambers. Air from the dinosaur's lungs was blown through the tubes (shown in pink) which acted as resonators producing a low note.

close together. Lake beach deposits, sandy river banks, and islands in lakes were also popular — all places where nest-digging is comparatively easy.

Dinosaur eggs come in a variety of shapes and sizes; round — up to the size of a football, elliptical and elongate — up to 60 centimetres long! However, even the largest dinosaurs laid comparatively small eggs because the shell must be thin enough to allow oxygen to pass through the pores in its surface and reach the developing embryo inside, and for the baby inside to break out. Eggs have their own scientific names and scientists can identify 'egg families'. But the identity of egg-layers is a different matter, and can be ascertained only when an identifiable embryo is found within the egg, or hatchlings are found clearly associated with eggshells. The first definite association of football-sized eggs with sauropod dinosaurs came with the discovery in 1998 of a huge egg site in Patagonia, Argentina, where thousands of eggs were distributed over an area of over 1 km^2. Bones, tiny 2 millimetre-long pencil-like teeth, and even the skin of unhatched babies, confirmed that the eggs belonged to sauropods, probably titanosaurs. The Gobi Desert nests and eggs were thought to belong to *Protoceratops*, until

EGG HORDE. These Chinese eggs, including whole or part clutches, show the huge variety of shapes and sizes that dinosaurs laid. They were just some of the 3000 eggs confiscated from smugglers by Chinese customs officers in 1993.

'Big Momma'. This spectacular fossil from the Gobi Desert proved conclusively that *Oviraptor* sat on its nest of eggs, squatting down like a bird with its forelimbs outstretched.

two exciting discoveries in 1993, of *Oviraptor* crouching on the same kind of nest. Laid by female *Oviraptor*, the nests consisted of up to twenty elliptical eggs laid upright in layered circles. To achieve this the mother must have first scraped a shallow hole about 1 metre wide and then crouched over it, gradually turning in a tight circle as each new egg was laid. Sand was pushed around the eggs to keep them upright, with the larger end at the top to allow the hatchling to escape. The discovery of *Oviraptor* preserved on top of its own nest of eggs, with its hind legs neatly folded and the eggs gathered between its forelimbs, was proof that small dinosaurs did sit on their nests. But was the adult protecting its eggs from sun, sand or predators, or was it actively brooding its eggs – keeping them at a constant temperature important for the developing embryos inside? The fossil evidence does not stretch that far, but the latter is a very reasonable interpretation, bearing in mind that some small meat-eating dinosaurs had feathers, as we shall see in Chapter 10. So the adult *Oviraptor* was probably doing exactly what birds do today — brooding — unaware that it was about to be engulfed by a sudden slumping of the sand dunes just behind its nest site. *Oviraptor* was less than 2 metres long, small enough to sit on its nest without crushing its eggs. Clearly, there was no question of most dinosaurs sitting on these fragile objects to incubate them, so they may have been covered with sand or relied on the surrounding air temperature being high enough to keep the uncovered eggs warm.

Helpless hatchlings. *Overleaf:* The *Maiasaura* young were too small and weak to leave the nest after hatching. They remained nest bound for many weeks and were fed and looked after by the adults. This enabled them to grow rapidly but left them vulnerable to attack from predators entering the nesting colony.

NESTING COLONIES

The first indication that some dinosaurs might have nested in colonies came after 1979 when a number of *Troodon* nests were found close together on two sites in Montana, USA. Ten years of painstaking excavation at these sites, and the study of the complete fossil record, has allowed John Horner and his team to build up a much fuller picture of dinosaur nesting behaviour.

Both sites were probably small islands off the shore of a shallow lake. Trees grew around the lake and the islands were covered with small plants. Several *Troodon* nests were found with eggs laid in clutches of 12 or 24, laid two at a time, narrow end down into the soil and tightly placed but not touching. *Troodon* constructed a nest by digging a bowl-shaped depression in flood plain

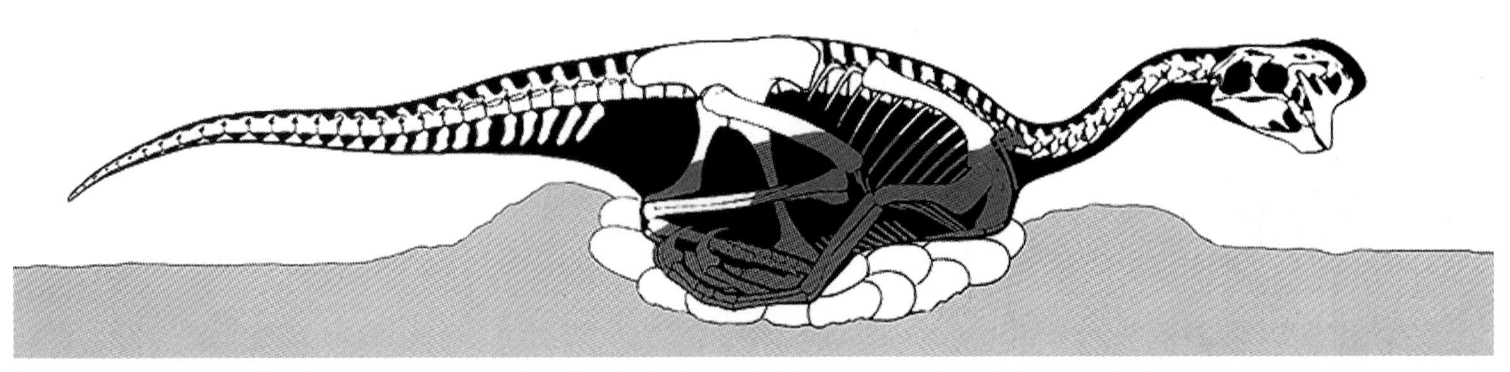

Sitting comfortably. A nesting *Oviraptor* reconstructed brooding its clutch of up to 22 eggs arranged in neat circles in a shallow nest bowl.

soils and using the loose earth to build a nest rim. The lack of any covering sediment or vegetation in the nest suggests that it probably brooded its eggs like *Oviraptor*.

A key feature of the *Troodon* nests was that the shells of the hatched eggs were found intact except where the top part was split off by the escaping young. This strongly suggests that the young *Troodon* walked away from the nest very soon after hatching, possibly under the protection of their mother, just as hens do. A well-developed *Troodon* young, found still inside its egg, had fully formed leg bones and joints, and would easily have been able to carry its own weight. The discovery of several half grown juvenile skeletons suggests that the hatchlings, although mobile, probably stayed near the nesting site for some while.

INSIDE THE EGG. Fragments of *Oviraptor* eggshell show the surface ornament which prevented blockage of the pores, allowing gas exchange through the shell.

Minute examination of the areas surrounding the *Troodon* nests gave a further insight into the balanced ecological system the dinosaurs were part of, and shows that the nesting colony supported a number of scavengers and predators. There are fossilized skeletons of varanid lizards. Modern varanid lizards frequently prey on the nests of crocodiles and scrub fowl, so their prehistoric ancestors probably invaded the *Troodon* nests. Fossil bones of small mammals suggest that they too may have eaten eggs or vulnerable hatchlings on the island. A large number of fossilized insect pupa cases or cocoons were also found. Modern beetles and insects are attracted to nests because the egg fluids and dead young are an easy and rich source of food, so the prehistoric beetles were probably scavenging in the same way. One egg was actually found with the hatchling inside having been eaten away by insects while still in the shell. Finally, there were discoveries of the fossil teeth of the much larger meat-eater *Albertosaurus*, a very likely predator on such a choice hunting ground.

CRACKED OPEN. A hatched *Troodon* egg; the top third was split off by the hatching young.

Knowledge of dinosaur nesting habits took a further leap forward with the discovery, again by John Horner, of another nesting colony in Montana. This one belonged to the plant-eating *Maiasaura* and turned out to have forty nests covering 1 hectare of ground.

Maiasaura, whose name means 'good mother lizard', was an eight metre-long hadrosaur. It built its nesting colonies in dry upland areas, presumably to avoid the danger of flooding and to have a commanding look-out position. This behaviour is similar to that of colonial birds such as terns who also rely on the colony to protect them from predators. *Maiasaura* returned to the same nesting site every breeding season and the nests were refurbished and used again. The nests themselves were about 2 metres in diameter in the form of a mud and stone walled basin lined with vegetation.

They were spaced one dinosaur length apart so that the parents could come and go without getting in the way of neighbours. Clutches of up to 25 eggs were laid in the nests and arranged in circles so that no egg touched another.

Maiasaura used a different method of incubation to *Troodon*, piling up vegetation, perhaps mixed with sand, on top of the nest and allowing the 'compost heap' to keep the eggs warm. Modern scrub fowl do this, removing and adding plants to keep the temperature constant. The striking difference between the two dinosaurs is that the *Maiasaura* young stayed in the nest for several weeks being fed by their parents before venturing out into the wider world. We can tell this because of the many fragments of trampled eggshell found in the nests and the remains of what appears to be regurgitated leaves and berries provided by the parent dinosaurs. The newly hatched *Maiasaura* measured around 30 centimetres long. Their leg bones and joints were not fully formed and it was not until they reached a length of about 1.5 metres that they left the nest. This growth would have been very rapid since staying still and being brought food by your parents is a very effective way to get big quickly (ask any baby starling). By the end of their first year the *Maiasaura* could have reached 2.5 metres or more in length and would be ready to migrate with their parents to lowland pastures. How long they stayed together as a family group within the herd is not clear. Growth studies of *Maiasaura*, detailed in Chapter 6, estimate that individuals probably reached breeding age six years or less. In terms of size, the nearest modern equivalent is a baby elephant which will stay with its mother for 15 years or more before becoming fully mature and independent.

The nesting behaviours of *Troodon* and *Maiasaura* give the clearest indication so far about the social development of dinosaurs. They mirror exactly the two kinds of nesting found among modern birds — 'precocial' where babies leave the nest as soon as they hatch, and 'altrical' where the helpless young remain in the nest. With *Maiasaura*, the large nesting communities suggest further social development. These dinosaurs must have had some kind of homing instinct, like swallows or pigeons, that guided them back to the same breeding ground year after year. Nest construction must have been instinctive knowledge, but the ability to construct a complex colony in co-operation with others suggests quite high degrees of communication skills. And once the *Maiasaura* young were moving about, the parents must have had some means of recognizing their own babies from among the many others in the breeding colony. Did they do this by smell? By special sounds? By distinguishing marks or colours? These questions are probably unanswerable, but they do show how socially advanced scientists now believe some dinosaurs to have been.

GOOD MOTHER LIZARD: *Maiasaura* herds returned to the same nesting grounds year after year. Nests were built one *Maiasaura* length apart. Up to 25 eggs were laid in each nest, and the hatchlings were fed by their parents. *Maiasaura*'s behaviour gives an insight into how dinosaurs operated as social animals.

LIVING A

6

Fossil remains tell us a good deal about dinosaurs' bones and teeth, but much less about the rest of their physical make-up. How did they see and hear? What colour were they? Were

NIMALS 6

they warm- or cold-blooded? Skilled research and comparison with modern animals give some insight into the physiology of dinosaurs but many fundamental questions are still unanswered.

The popular image of dinosaurs is as huge, amazingly shaped creatures more terrifying than the dragons of legend or fairytale. In fact, they were quite ordinary living animals with bodies that had to perform the same basic functions as any dog or tortoise or human being. To understand dinosaurs as real live animals we must look beyond the 'Did-you-know? Amazing Facts!' about size or weight or length of teeth and examine their biological make-up.

All dinosaurs were built around a bony scaffolding that had the same basic components — skull, spine, ribs, shoulders, hips, legs and tail. The hips tell us which of the two main groups a dinosaur belongs to. To determine this when looking at a dinosaur skeleton, first find the pelvis or hip girdle, where the back legs join the spine. The pelvis is made up of three bones called the ilium, the ischium and the pubis. If the ischium and pubis radiate out from the ilium in different directions, with the pubis pointing forward between the legs, then the dinosaur is a saurischian (lizard-hipped). All meat-eating dinosaurs belonged to this group, as did all the plant-eating sauropods and their relatives. Modern lizards and crocodiles have a similar hip structure.

Dinosaurs in the other group have quite different hips, similar superficially to those of modern birds, and hence are called ornithischians (bird-hipped). Here the pubis is swung right back to lie alongside the ischium, pointing backwards down the tail. All ornithischians were plant-eaters and they account for the vast majority of dinosaur fossils discovered. Despite their 'bird-hipped' group name, however, there is no evolutionary link between this group of dinosaurs and birds.

(*Previous spread*) STAYING ALIVE. In the late Cretaceous Period in North America three groups of hadrosaurs — *Parasaurolophus*, *Corythosaurus* and *Lambeosaurus* — sense danger from an approaching predator.

BONES AND MUSCLES

Having said that dinosaurs all had the same basic bony scaffolding, the bones themselves show many variations. For the vast plant-eaters, strength was the main requirement and their leg bones were huge and solid, capable of carrying enormous loads. At the same time they developed an ingenious system for reducing the weight of other bones without any loss of strength. The vertebrae of the sauropod *Ornithopsis*, for example, had great hollows, called pleurocoels, scooped out of them. Yet the remaining structure was just as strong as a solid lump of bone despite being nearly 50 per cent lighter.

Fossilized bones can yield surprisingly detailed information about the once-living animal. Soft tissues have been preserved in a specimen of *Tyrannosaurus rex* that was about 18 years old when it died. Its bones were mummified and embedded in soft sandstone, which preserved transparent, flexible, hollow blood vessels containing probable blood corpuscles. The bone also retained traces of its original organic elastic matrix called collagen. The structures identified by Mary Schweitzer compare closely to the equivalent tissues in a modern ostrich. This is not so surprising since, as we shall see in Chapter 10, birds are descended from theropod dinosaurs. But the preservation of such tissues in fossil bone after 65 million years is quite remarkable. Even more so was Schweitzer's discovery of medullary bone in the same *Tyrannosaurus rex*.

BONE ENGINEERING. This vertebra from the middle of the backbone of *Ornithopsis*, a large four-legged plant-eater related to *Diplodocus*, has extensive hollows (pleurocoels) to reduce its weight. The bone is virtually a series of struts joined together at critical points to give maximum strength for minimum weight.

THEM BONES. *Top:* Speed was important to *Dryosaurus*, a small unarmed plant-eater. Thin-walled, hollow bones, similar to the modern gazelle, gave strength without adding weight to the skeleton. *Bottom:* Solid pillar-like limb bones supported the 20–30 tonne *Apatosaurus*. This thigh bone is 1.5 metres long.

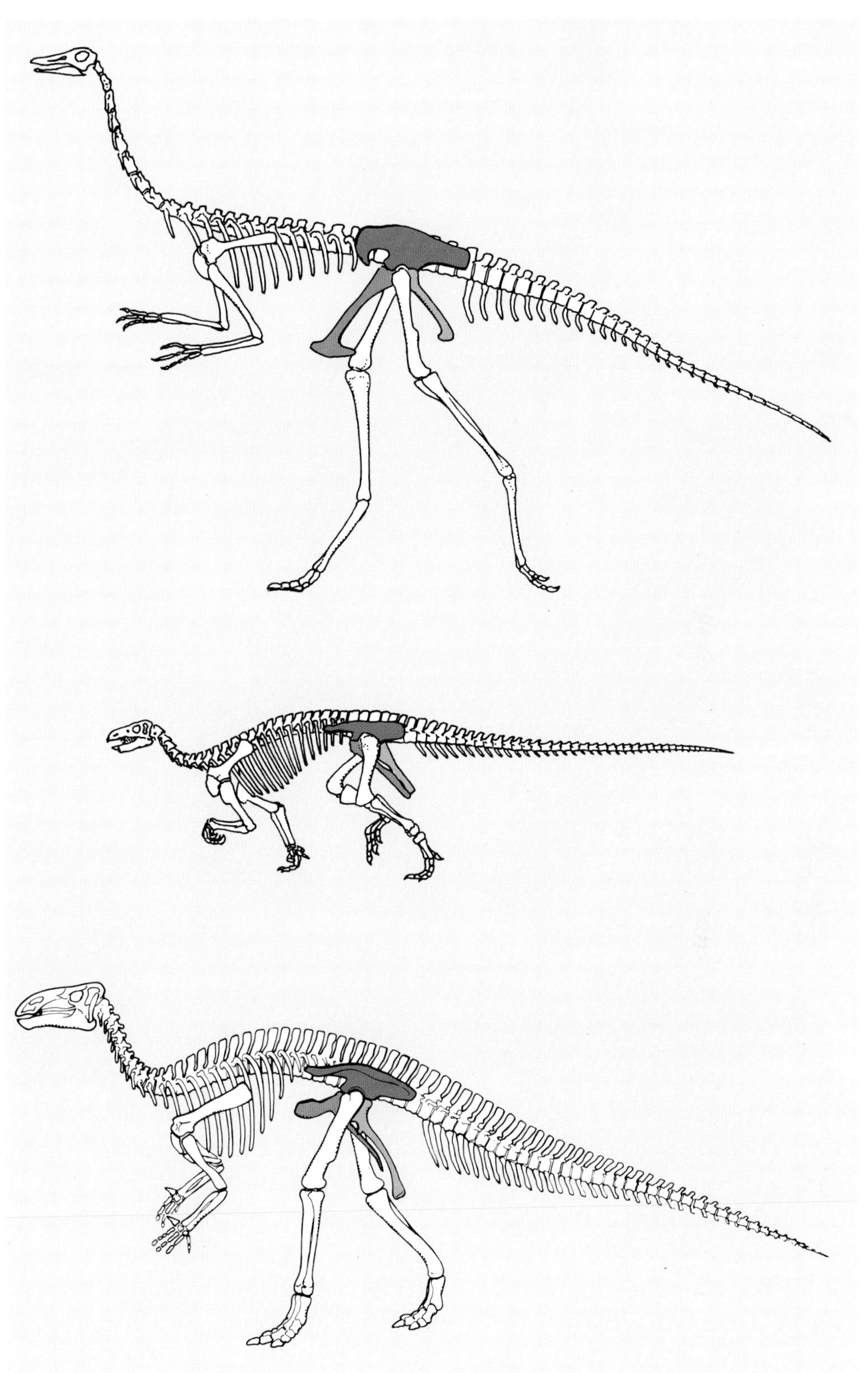

HIP TALK. The pelvis consists of three bones on each side — ilium (red), ischium (yellow) and pubis (green). *Top:* In saurischian dinosaurs the ischium and pubis point in different directions. *Middle:* Early ornithischians, like *Scelidosaurus* have the ischium and pubis lying together and pointing backwards towards the tail. *Bottom:* Later ornithischians, like *Iguanodon*, developed an extra forward-facing prong on the pubis. This does not make them saurischians.

Medullary bone is characteristic of ovulating birds and is formed when calcium is drawn from the bones to make eggshells. So here is a way to tell the sex of at least one *Tyrannosaurus rex*! It also confirms that dinosaur and bird reproductive physiology was identical.

The proteins that control and organize the laying down of a form of calcium phosphate, called 'apatite', in crystals in the elastic collagen matrix to form bone, have been detected in *Iguanodon* bone by a team led by Graham Embery and Angela Milner. The remains of these proteins are even older, they survived 125 million years by being locked deeply inside the apatite crystals of the bone. Both these studies have opened up the

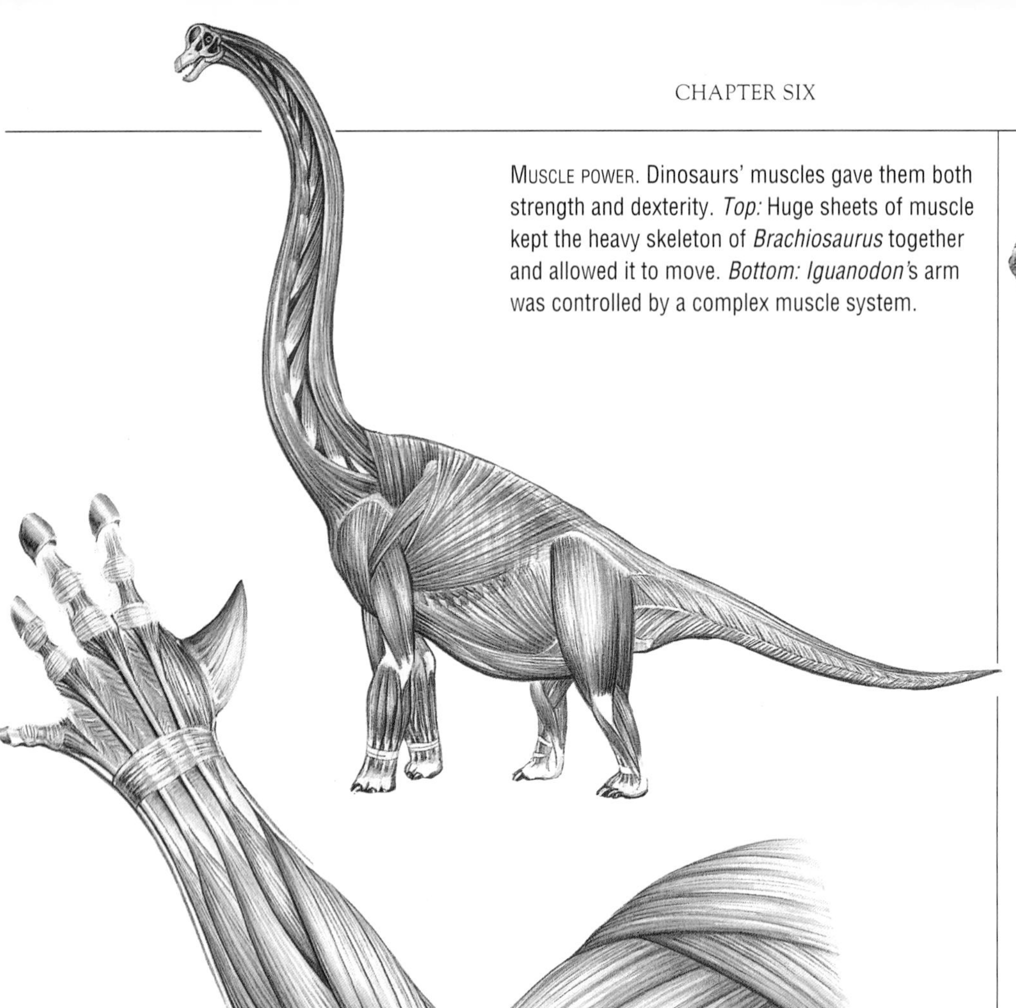

MUSCLE POWER. Dinosaurs' muscles gave them both strength and dexterity. *Top:* Huge sheets of muscle kept the heavy skeleton of *Brachiosaurus* together and allowed it to move. *Bottom: Iguanodon*'s arm was controlled by a complex muscle system.

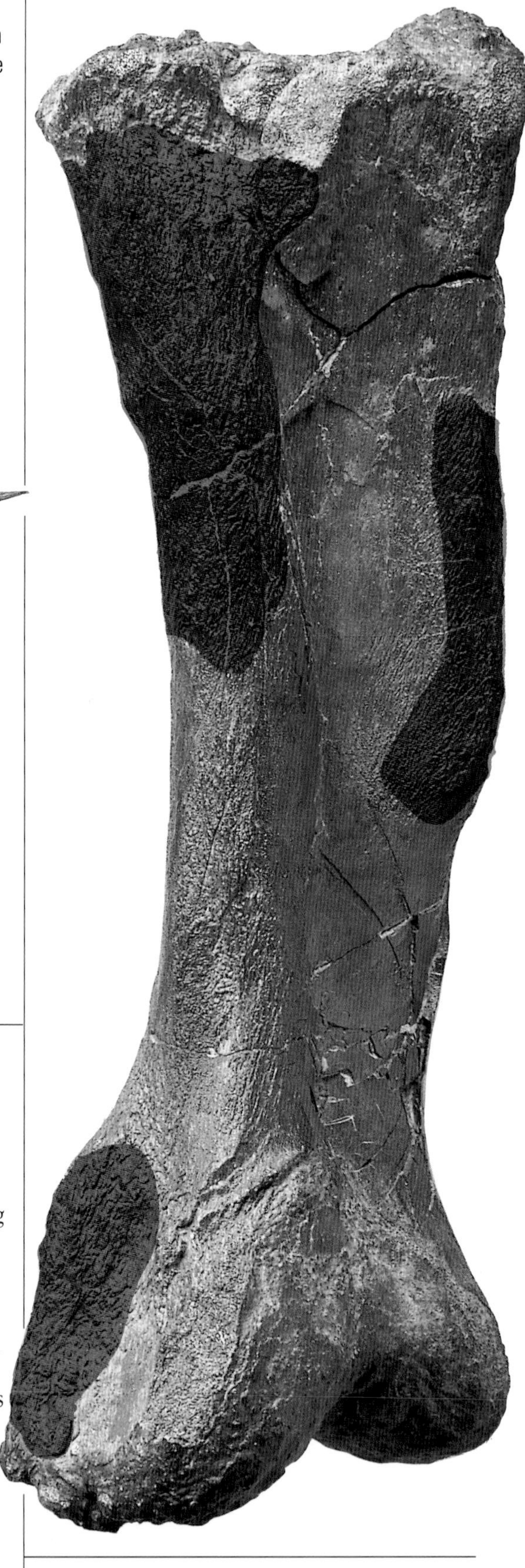

BONE SCARS. The rough patches (as shaded) on this *Iguanodon* metatarsal (long bone of foot) show where the muscles and tendons were attached.

exciting prospect of dinosaur studies at the cellular, or even molecular, level.

In life, each dinosaur's skeleton was held together by a network of ligaments, muscles and tendons, just as our bodies are. Some fossil bones actually have 'muscle scars' (roughened patches where the muscles were attached) and from these it is possible to work out the size and position of some main controlling muscles. Anatomists can further interpret the unique structure of each skeleton to calculate where the muscles were located, how the dinosaur moved and the overall shape of its body. We also know that, like modern reptiles and birds, dinosaurs had relatively few muscles in their faces and were probably rather expressionless. The fully restored head models you may see in some museums look very similar to dinosaur skulls, being basically just skin over bone with very little extra padding in between.

Yet even here, at this most basic level of building on the bony framework, there is plenty of room for doubt. The legs of a large plant-eater like *Diplodocus*, for example, must have been moved by huge sheets and bunches of muscle of which there is no trace in the fossilized remains. The expanding jaws of *Tyrannosaurus* were controlled by a set of muscles and tendons that interacted with each other in highly complex ways. How far could *Tenontosaurus* move its tail up and down, and how far from side to side? Did *Albertosaurus* have a deep muscular throat like a bulldog or a high ridge of muscle at the back of its neck? No one knowns for certain and, while modern animals can sometimes provide clues, this is certainly not hard evidence. Fundamentally, the amount and relative proportions of muscle that you think each dinosaur carried is linked with how you believe it moved and lived. It is fascinating to see the different illustrations of the same dinosaur over the years as views have changed about its way of life. Early pictures of *Tyrannosaurus*, for example, show it draped in quite puny muscles consistent with the sluggish animal it was thought to be. More recent ideas of *Tyrannosaurus* as an active hunter have produced illustrations which show a huge, muscular animal. One of the joys, and also the hazards of dinosaur study, is how much is left open to interpretation.

SKIN AND ARMOUR

Over the top of their skeleton and supporting muscles, each dinosaur wore a skin and a variety of scales, plates, spines, claws and horns. As described in Chapter 4 (Attack and Defence), fossilized skin is very rare but we know that it is made up of thousands of small bony plates (called osteoderms) embedded in the skin, rather like the covering of an alligator or a lizard.

On the heavily armoured ankylosaurs the bigger plates and scales formed a defensive barrier to prevent attack. On *Stegosaurus*, the large chevron-shaped back plates are no longer thought to be primarily for defence (they would have been useless for protecting the most vulnerable belly parts in any case) but rather to be part of a body heating and cooling system. *Stegosaurus* pumped blood up through the skin that covered the plates and, by standing side on to the Sun, could use them for rapid warming. If it was too hot, it simply had to move into the shade or face the Sun so that the minimum plate area was exposed. A similar feature has been found in a group of theropod dinosaurs called spinosaurids and in a relative of *Iguanodon* called *Ouranosaurus*. These dinosaurs had elongated back spines up to 1.8 metres long which may have supported a huge sail-like fin. Such fins would have provided an enormous surface area for heating or cooling blood.

Claws and horns were supported by a bony core and sheathed in a covering of tough keratin, the same substance that your fingernails are made of. Claw and horn cores have a groove running along

SHARP ENCOUNTERS. The strong horny covering of *Triceratops*' bony horn cores made them fearsome weapons. Even a large hunter like *Tyrannosaurus* could easily sustain a fatal injury from a single thrust.

BRIGHT EYES. The large eyes and brain of *Troodon* suggest it was intelligent and active.

one or both sides that was used to anchor the horn covering. Plant-eaters' claws were usually blunt or hoof-like but meat-eaters' claws were sharply pointed. We have looked already at dinosaur colouring but it is worth repeating here that any ideas about the colour of dinosaurs' skin is speculation rather than fact. Even more controversial is the question of whether some dinosaurs had feathers. This remarkable idea was first suggested in the late 1970s by the Russian scientist S. M. Kurzanov when studying the small meat-eating *Avimimus*.

Feathers would have been useful not for flight but for insulation and display, so perhaps some small meat-eaters had them too? New fossil sites in the early Cretaceous of China have yielded stunning and crucial evidence to answer that intriguing question — several different small meat-eating dinosaurs complete with feather or feather-like body coverings. These are described in Chapter 10. So, it is quite legitimate to illustrate such dinosaurs with downy feathers, neck plumage, arm and even tail feathers, as now routinely depicted by many artists and model-makers. Such ideas are now widely accepted and most dinosaur experts are convinced that there is a direct link between dinosaurs and birds.

CLEVER OR DUMB?

Were dinosaurs stupid? Describing *Brontosaurus* (now called *Apatosaurus*) in 1883, the great American palaeontologist Othniel Marsh stated that its tiny brain proved it to be a "stupid, slow moving reptile", words which probably still sum up the general view of all dinosaurs' mental abilities. The facts tell a rather different story, as we can see when we start to look in detail at the dinosaurs' senses and the brains that controlled them.

Much work has been undertaken recently to determine the size of dinosaurs' brains relative to their body weight. This is fairly easy to do where there is a well-preserved skull, since all dinosaurs had brains enclosed in a tough bony case. If you measure the volume of this braincase inside the skull and take into account the variation in the percentage of the cavity occupied by the brain (not all the space was actually filled with brain), this will give you the actual brain size.

There is no doubt that some dinosaurs got by with very little by way of brain. *Stegosaurus* weighed 3.3 tonnes and had just 60 grammes of brain, thirty times less than an elephant of the same weight. The ratio of brain to body weight in the large sauropods is even more dramatic, 100,000:1. Yet this is hardly surprising. Everything we know about these dinosaurs suggests that they led lives that demanded little brain work. *Brachiosaurus* did not have to hunt its food or evade predators, both activities that need brain power. *Stegosaurus* may well have lived in herds or family groups, but was not crucially dependent for survival on herd communication and rapid reaction, as were the unarmoured (and therefore far more intelligent) hadrosaurs. Simple lives demand less control and co-ordination.

What we should expect to find is increasing brain size and complexity in line with the demands made by the dinosaurs' lifestyle — slow plant-eaters at the bottom, herders and pack hunters in the middle and fast-running stalker hunters at the top. This is actually what Professor James Hopson of the University of Chicago, USA, discovered when he compared dinosaur brain and body sizes to determine the 'braininess' of different groups. According to his research the 'league table' of intelligence runs, from the bottom up: sauropods, ankylosaurs, stegosaurs, ceratopians, ornithopods, carnosaurs and coelurosaurs.

The structure of dinosaur skulls also provides clues to the relative size, importance and complexity of their sense organs, and hence their brains. For example, large forward-pointing eye-sockets suggest a sight-dominant animal; large spaces in the snout could mean the sense of smell was important.

Most dinosaurs had monocular vision with eyes set at the sides of their heads, and very little overlap between the left and right fields of view. This gave them a good wide-angled view of their surroundings but no real ability to judge distances. For this, front-facing eyes are required and considerable brain power to process and interpret the visual information. There is evidence that some of the big meat-eaters like *Tyrannosaurus* had at least some binocular overlap, but the dinosaurs which developed fully binocular vision were the smaller meat-eaters. Their brains were particularly well developed in the area that controlled the co-ordination between running, hand movement and visual information about moving objects. This was the key to the hunting success of dinosaurs like *Troodon* that stalked and chased scuttling mammals and lizards. It had exceptionally large eyes, based on the size of the eye sockets, and the biggest brain of any dinosaur relative to its body size — certainly up to the level of some modern birds and mammals.

Just how dinosaurs saw the world we will never know. Some reptiles see in colour as do most birds, which have the keenest of vision of all living animals. Hawks, for example, have up to eight times more visual cells packed into the most sensitive spot of their eyes than we do, allowing them to recognize shapes that would only be blurs to us. The relative size of birds' eyeballs is also huge, in some cases taking up more space than the brain itself (an ostrich's eye is the size of a tennis ball). All the evidence suggests that sight was also a critical sense for the dinosaurs, ranging from the limited low-level view of *Protoceratops* or *Tuojiangosaurus*, through the wide-angled horizon-scanning lookout ability of the hadrosaurs or iguanodontids, up to the keen hunter's vision of *Troodon*.

BRAIN BOX. In this *Iguanodon* braincase, the brain tissue decayed before fossilization, leaving a hollow space. A silicone rubber cast shows the approximate shape of the brain as it would have fitted in the braincase.

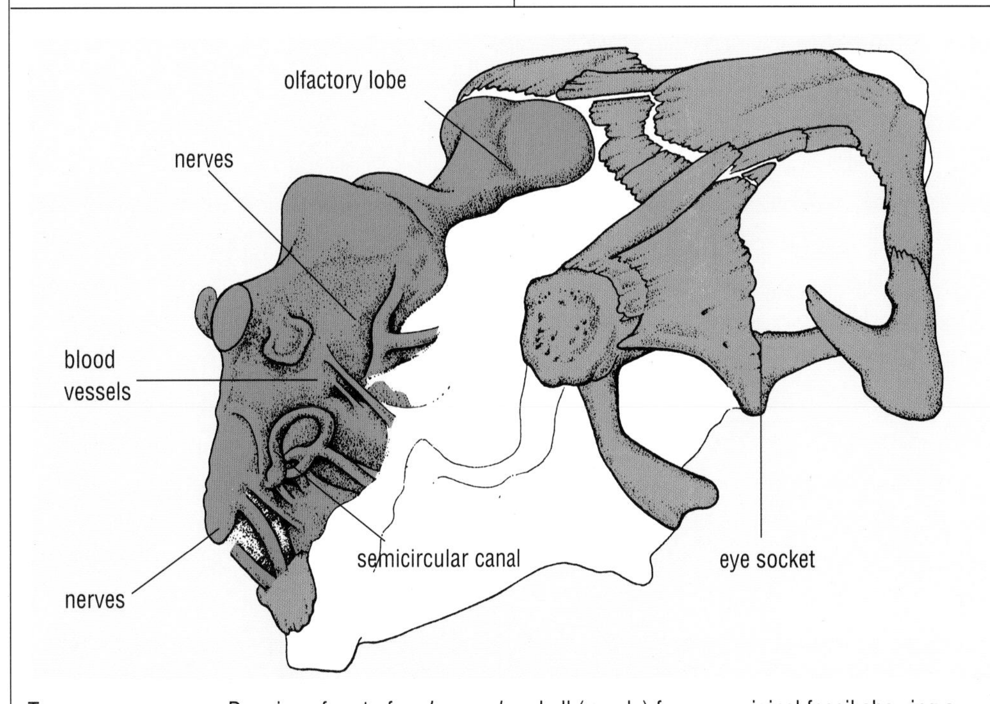

TASTE, SMELL, BALANCE. Drawing of part of an *Iguanodon* skull (purple) from an original fossil showing a preserved brain (orange, the equivalent of the silicone cast in the figure above) prepared from the surrounding rock (white). The large olfactory lobe at the front suggests it had a well-developed sense of smell. Nerves and blood-vessels to the eyes, face and ears can be seen. Even the semicircular canal inside the ear is preserved. This would have given *Iguanodon* its sense of balance.

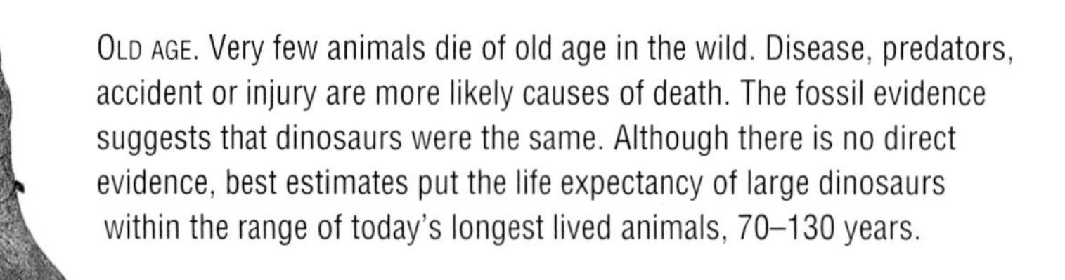

OLD AGE. Very few animals die of old age in the wild. Disease, predators, accident or injury are more likely causes of death. The fossil evidence suggests that dinosaurs were the same. Although there is no direct evidence, best estimates put the life expectancy of large dinosaurs within the range of today's longest lived animals, 70–130 years.

Left: An *Iguanodon* foot showing hard frills of arthritic growth on the toe joints. *Right:* This *Iguanodon* hip bone was fractured during its life and healed slightly out of alignment.

Dinosaurs did not have exterior ear flaps like mammals, but heard through holes set far back in the head behind their eyes. Today's reptiles and birds have a similar arrangement with a slit-like external ear opening leading to the hearing organs in the brain. For birds in particular, hearing is an important part of their social behaviour, closely linked with their ability to communicate by song. A male bird can have several hundred songs in its repertoire each with a particular meaning understood by other members of the same species. It was probably herding dinosaurs, with their strong need to communicate, that had the most acute sense of hearing. Indeed, hadrosaurs display the only real evidence so far discovered of being able to make noises, with a variety of nasal trumpets and air sacs. Such equipment could not produce the complex trills, tiny pauses and extra-high notes achieved by birds, but they could certainly have allowed the development of a basic hadrosaur language for warning and directing the herd, as well as playing a crucial role at mating time. Co-operative hunters like *Deinonychus* would also have needed to communicate when stalking or bringing down a victim. Judging by the behaviour of modern pack animals, however, this was probably a much more visual than noise-based language (lying in ambush or running at full pelt are hardly ideal moments to bark orders).

Dinosaurs, like many modern land animals, would probably have made noises as and when the occasion demanded — to defend territory, attract a mate, see off an enemy or warn of danger. If we were able to play back a 'sound track' of the dinosaur era, we would probably hear a good range of snuffles, squeaks, grunts, chirps, snarls, roars and howls.

The senses of taste and smell are closely linked in all animals, being controlled by the same part of the brain. They are used by hunters and hunted to keep track of each other, by some plant-eaters to discriminate between attractive and unsuitable food and, in many species, as part of the mating process. It is hard to believe that the big meat-eaters like *Allosaurus* did much tasting of the meat they bolted down their huge throats. Their tongues were probably simple and rough, designed to position chunks of meat for swallowing as fast as possible. Plant-eaters, particularly those that chewed their food, would have needed more mobile tongues to move the wads of vegetation to the correct position for chopping and grinding. One of the very rare fossil brains that has been preserved, that of an *Iguanodon*, shows well-developed olfactory lobes (the section of the brain dealing with smell and taste) at the front of the brain. *Iguanodon* certainly had a large, broad snout to accommodate nostrils and sense organs, so perhaps this dinosaur at least had a keen sense of smell and relished the taste of its food. Smell is important to meat-eaters too. Modern smell-oriented turkey vultures can detect carrion from several kilometres away and they have large olfactory lobes or bulbs compared to black vultures which pinpoint their food by eye. A CT scan of the skull of 'Sue', a *Tyrannosaurus* specimen, revealed very large olfactory bulbs at the front of the brain. As we saw in Chapter 3, tyrannosaurs were likely to have been at least part-time scavengers, so a good sense of smell would be important in the business of locating the next meal.

The sense of touch cannot have been very well developed in any dinosaur, simply because of the nature of their

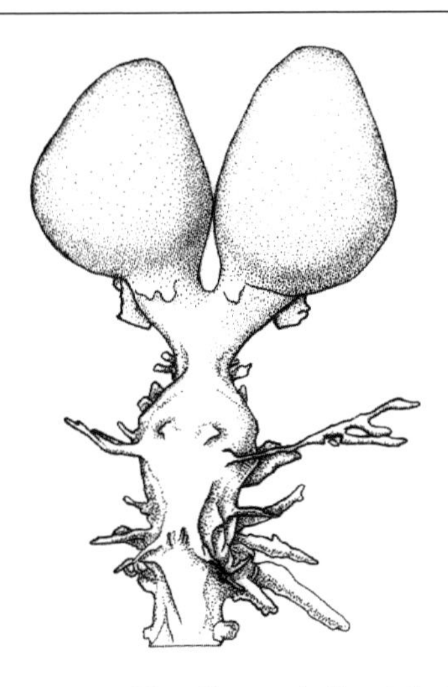

SMELL DETECTORS. The olfactory bulbs at the front of the brain process incoming smells. The huge ones in *Tyrannosaurus* suggest that it received powerful scent messages important for locating its food.

skin. Bony plates embedded in a thick hide are not designed for sensitivity and dinosaurs could have scraped and blundered their way past trees and rocks without feeling very much at all. Alligators have sensitive skin spots on their snouts and necks which they rub against each other at mating times. Is it too fanciful to see a pair of *Tyrannosaurus* doing the same? Or a couple of *Diplodocus* with necks lovingly entwined having a quick nuzzle in the high branches? Looking at dinosaur senses does not produce a definitive answer as to whether they were clever or stupid. Indeed, these terms are not really relevant since they imply that dinosaurs would have somehow been 'better off' if they had more human-like intelligence. The fact is that each species of dinosaur had all the brain power it needed to succeed at its own way of life for many millions of years. If humans achieve a fraction of this, we shall be doing well.

THE BLOOD QUESTION

Were dinosaurs warm- or cold-blooded? Dinosaur experts care passionately about this subject, many holding radically opposed and irreconcilable views. To understand why this debate has raged so fiercely for the last twenty years we must start by looking at the nature of warm- and cold-bloodedness itself.

Animals function most efficiently if their body temperature is fairly constant. The chemical reactions that take place in a body work best at set temperatures and if these go up or down too far, the body ceases to operate properly. So-called cold-blooded animals, like lizards and snakes, control their body temperature by their behaviour, moving in and out of the sun during the day. This is called the ectothermic ('outside heat') method. Warm-blooded animals (birds and mammals) convert food energy into body heat, the endothermic ('inside heat') method. To cool off, endothermic animals sweat, pant, wallow in water or, in the case of an elephant, flap their ears to cool down the blood that flows through them.

There are advantages and disadvantages to both systems. A warm-blooded dog burns off its food energy very rapidly and needs to eat nearly ten times more than a cold-blooded lizard of the same size. On the other hand, a lizard must spend several hours a day basking in the sun to warm up and cannot operate effectively at night or when the surrounding temperature drops. More importantly, as far as the dinosaurs are concerned, warm-blooded animals can support much larger brains and more active lifestyles than cold-blooded ones because brains in particular need constant temperatures to function efficiently. So the issue of warm or cold blood actually determines the whole way you look at dinosaurs — sluggish and dim, or active and intelligent more like today's mammals and birds. No wonder dinosaur experts think it is so important. Evidence quoted by each side can be complex and controversial but the main issues are easy to understand.

Many of the big dinosaurs held their heads high above their bodies — *Tyrannosaurus* and *Iguanodon* are examples and *Brachiosaurus* an

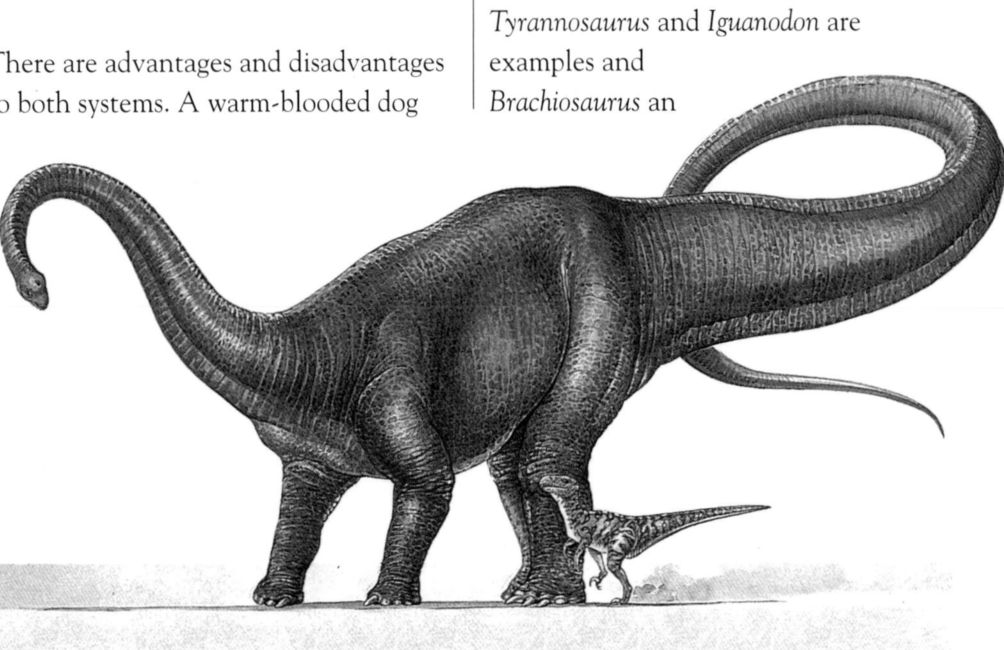

LARGE AND SMALL. Dinosaurs had very varied lifestyles ranging from huge slow-moving plant-eaters to smaller, active hunters. Did they also have a range of different warm- and cold-blooded metabolisms?

extreme case. To pump blood up to their brains would have required high pressure, far higher than the delicate blood-vessels in their lungs could withstand. Warm-blooded birds and mammals deal with this problem by having two blood circuits — a low pressure one to pick up oxygen and release carbon dioxide in the lungs, and a high pressure one to transport oxygen and nutrients to the rest of the body. Their hearts are internally divided and run the two circuits side by side. Lofty dinosaurs would also have needed a divided heart and, say some scientists, this suggests their bodies worked like those of warm-blooded birds and mammals. It is this final link in the logic that is controversial. Some dinosaurs must indeed have had a divided heart, or something very like it to get the blood to their heads, but they would not necessarily need to be fully warm-blooded to achieve this. Modern alligators have a functionally divided heart and remain cold-blooded. There is a great energy saving advantage in using the sun to warm up rather than having to go and find food to do the same job. In evolutionary terms, dinosaurs may have retained this advantage as long as possible, perhaps combining a version of the warm-blooded heart and lung system with cold-blooded ways to control body temperature.

The huge size of some dinosaurs and active lifestyles of others have also been put forward as arguments for warm-bloodedness. A 30 tonne sauropod could never warm up in the Sun, it is said, because its surface area was so small compared to its volume. At the other end of the scale, a fast-running, leaping, back-claw slasher like *Deinonychus* could never have maintained such a high level of activity without having a warm-blooded animal's ability to generate its own body heat.

Yet these arguments may seem less convincing when you remember the unchanging warm climate that persisted during most of the dinosaur era. Under these conditions, the large body and small surface area of a huge plant-eater like *Diplodocus* would have prevented it cooling down, enabling it to maintain a constant body temperature. Indeed, with the heat generated by its fermenting tank stomach, the problem might well have been how to lose heat rather than keep it. As for *Deinonychus*, just because it could run and jump does not mean it did so very often. As a cold-blooded hunter it could afford to kill and eat far less often than if it were warm-blooded, leaving plenty of time to top up body heat levels by sun-basking. The most we can say is that, on balance, the intelligent pack-hunter's life led by *Deinonychus* points strongly towards it having a physiology similar to modern birds and mammals.

Microscopic study of thin sections of dinosaur bone has provided more clues, and controversy, about dinosaur

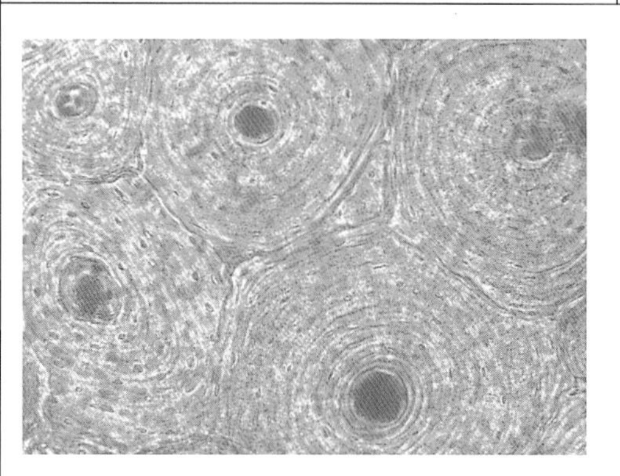

NEAR THE BONE. Haversian bone has many blood vessels with dense bony rings around them. Today's large warm-blooded mammals have this type of bone. This section is from the rib of a horse.

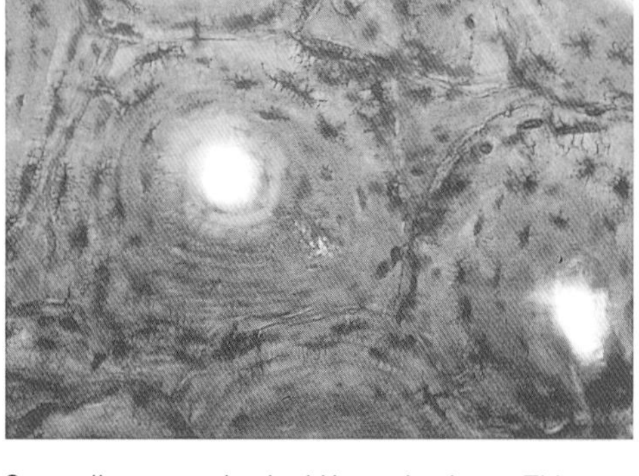

Some dinosaurs also had Haversian bone. This section of a *Baryonyx* rib shows the same bony ring structure as the horse. Does this mean that dinosaurs with Haversian bone were warm-blooded?

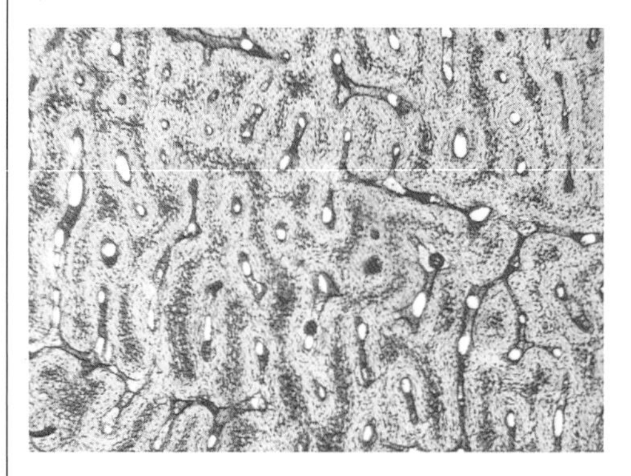

Dinosaurs, birds and mammals all have similar primary bone with many blood vessels. This type of bone, called fibro-lamellar bone (shown here from a sauropod limb bone), is the first to be formed during rapid growth.

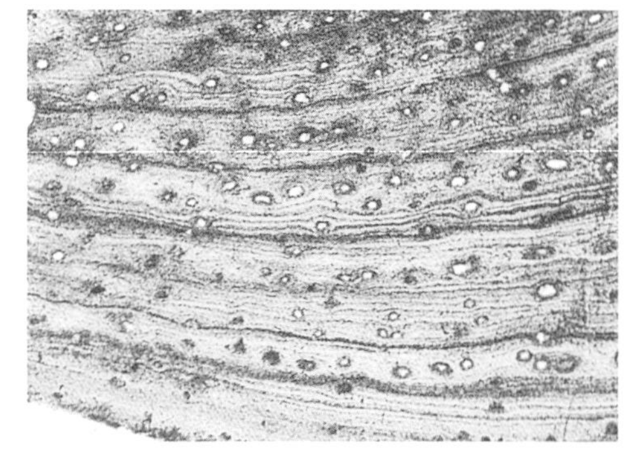

Modern cold-blooded reptiles show growth rings in their bones indicating that they grow at different rates. But some dinosaurs, as in this iguanodont limb bone, also show this feature.

So dinosaurs had all three kinds of bone, evidence of a physiology in between that of cold-blooded reptiles and warm-blooded mammals and birds.

physiology. Some dinosaur bone shows the same kind of microscopic structure as that of large modern mammals, which contains many blood vessels. This kind of bone (Haversian bone) is formed by 'remodelling' the bone tissues as the animal grows to provide more stress resistance and strength. It is accomplished by laying down dense tubes of bony tissue round each of the many blood vessels. Since dinosaurs had the same kind of bone this could mean that they were warm-blooded like mammals, say the hot-blooded dinosaur supporters. The discovery that small plant-eating hypsilophodonts lived in polar conditions in southern Australia in early Cretaceous times, and encountered some sub-zero temperatures, is another interesting point in the warm-blooded debate (see Chapter 8). However, detailed studies on many different dinosaurs, particularly by Dr Robin Reid, at the Queen's University of Belfast, Northern Ireland, have shown that the evidence is just not that simple.

In many dinosaurs, the bone formed first during rapid growth (the primary bone) is similar to that of birds and mammals with lots of blood vessels. This is an important difference between dinosaurs and modern reptiles which lay down their bone in seasonal growth rings a bit like tree rings. However, some dinosaurs show this kind of bone too and up to a point it can give some idea of the age of an individual, except that we do not know if one ring equals one year! So, dinosaurs have three types of bony tissue: primary bone, Haversian bone and growth ring bone, and it varies between bones and even within the same bone.

What does all this mean? The work of Dr Reid and others has shown that dinosaurs could grow very rapidly (as fast as birds and mammals) when they were young as they had a lot of primary bone tissue, especially in their legs. This tissue was sometimes replaced by secondary (Haversian) bone to give strength and stress resistance as it does in large mammals today. The seasonal growth rings support the idea that their growth slowed down later in life and became more reptile like. As Dr Reid suggests, dinosaurs were not quite like reptiles, birds or mammals, but had a uniquely dinosaurian physiology all of their own. Dr John Horner has recently led a detailed study of the bone tissues of *Maiasaura* and compared them to dinosaurs' nearest living relatives — crocodiles and birds. Newly hatched *Maiasaura* nestlings grew at very high rates indeed; later stages grew at high to moderately high rates that slowed to low or very low growth in adults which were 7–9 metres in length. These growth rates translate into estimates of a nesting period of one to two months with late juvenile size (3.5 metres long) reached in one to two years, and adult size reached in 6–8 years. Growth rings appear regularly by the sub-adult stage (4.7 metres long), showing that growth rates had slowed down. Sauropods, the biggest dinosaurs of all, show a similar pattern with continuous fast growth in young animals and a dramatic slowing down in individuals approaching maximum size. Dr Martin Sander has calculated that sexual maturity (marked by the end of the fastest growth period) was achieved at 40% of maximum size in *Brachiosaurus* and 70% in *Barosaurus*. In one individual of the sauropod *Janenschia*, Dr Sander has made a qualitative estimate of its life history based on unusual growth markers — 'polish lines' in the bone — it was sexually mature at about 11 years, grew only imperceptibly by the time it was 26 and died at around 38 years old. Was that typical for the largest dinosaurs?

Tyrannosaurus has been shown to have a highly unusual growth pattern by a research group led by Greg Erickson. It underwent a 'teenage' growth spurt between the ages of 15 and 20 years. Over about a four-year period, *Tyrannosaurus* put on weight at the astonishing rate of 2.07 kg per day and then virtually stopped growing at around 20 years old when it reached skeletal maturity and a weight of 5000 kg or more. The body mass estimate for the largest-known complete specimen, 'Sue' at the Field Museum in Chicago, is 5654 kg and 'she' died at age 28. *Tyrannosaurus*' growth rate had a bearing on its running speeds. It would have been incapable of fast running once it passed the 1000 kg mark, which corresponds to a juvenile aged about 13 years. Older animals became just too large and heavy to run fast, they simply could not have had enough leg muscle in relation to their body weight. John Hutchinson, a biomechanics expert, estimates a rough upper-end estimate of walking speed at about 18 km/h, but faster or slower speeds are not impossible. He argues that other calculations of a 'safe' top speed of 40 km/h are very questionable — 41% of *Tyrannosaurus*' body mass would need to have been leg muscles for it to run this fast. There is plenty more to be said but what it all amounts to is a definite 'maybe' about dinosaur warm- and cold-bloodedness.

WHERE DID

7

Some catastrophic event wiped out the dinosaurs. Was it an asteroid collision? Or a climate change brought about by massive volcanic eruptions? Buried under tonnes of

THEY GO? 7

rock, the evidence is elusive and controversial. Yet it all points to the dinosaurs' involvement in one of the most disastrous mass-extinctions in the Earth's history.

Why do animals die out? Every species evolves to exploit a particular environment, sometimes in a very specialized way. The koala bear, for example, is the only mammal that can survive on a diet of eucalyptus leaves, thanks to a unique digestive system and an extraordinarily long gut. While this diet means it is restricted to living in eucalyptus-growing areas, there is no competition for its food supply. The koala bears can have their eucalyptus trees to themselves, eating just 1 kilogramme of leaves each night and sleeping for twenty-two hours out of twenty-four.

On the other hand, beetles have adapted to thrive almost anywhere. By virtually any measurement they are one of today's most successful animal groups with at least half a million species so far described, distributed over virtually every part of the Earth's surface. Indeed, if beetles were allowed to develop unchecked for just a few months, the whole land surface of the planet would be covered with them. While beetles and koala bears have evolved quite differently, they both owe their success to the same cause — being adapted to do better in a particular part of the ecosystem than any other animal. Being well adapted to an environment is all very well, except for one thing — the environment itself is constantly changing. Extinction can happen when a species does not adapt quickly enough to meet these changes.

For over 160 million years dinosaurs were champion adapters. Although each individual species may only have survived for 2–3 million years, the dinosaur group as a whole was remarkably good at meeting any new conditions. As explained in Chapter 1, continents moved, sea levels rose and fell, volcanoes erupted, weather patterns changed and new plants appeared, but the dinosaurs thrived on it all. They were diverse, they lived on every continent and they totally dominated the land. Then, quite suddenly, they all disappeared. What happened? Almost everyone now accepts that dinosaurs were finally wiped out by a sudden, catastrophic event or events that affected the Earth's climate. This change was so dramatic and so fast that the dinosaurs had no chance to adapt before the new conditions killed them. What no one can say for certain is how the event came about.

THE END OF THE DINOSAUR ERA

This is what we do know. During the last few million years of the dinosaur era, a more seasonal climatic pattern of warm summers and cool winters developed, particularly at high latitudes. This was influenced by changes in the seas and oceans as the continents moved apart. The fossil record shows that in western North America, which has produced the only continuous late Cretaceous dinosaur fauna, it was at its most diverse about 76–74 million years ago, but by 67 million years ago a small number of very successful plant-eaters dominated the scene. These included a hadrosaur, *Edmontosaurus* and a ceratopian, *Triceratops*. Meat-eaters like *Tyrannosaurus* and *Troodon* hunted and scavenged on the fringes of the herds.

It now seems possible that the herds themselves migrated over much larger distances than scientists originally thought. Abundant remains of *Edmontosaurus*, together with bones of a horned dinosaur, *Pachyrhinosaurus*, and teeth of a large tyrannosaurid have been found on the North Slope of Alaska, which was well inside the Arctic Circle even 70–65 million years ago. This discovery, made in 1986, indicates that dinosaurs could live at high latitudes even though the climate was cool and there was a long period of darkness in winter (but no polar ice-caps as there are today). The associated plant remains tell us that the temperature ranged from an average 10–12 °C in the warmest month down to 2–4 °C during the coldest. Did the dinosaurs live there all year round, spending three months in cold, dark conditions? It seems very unlikely. A much more plausible explanation is that herds of plant-eaters, followed by their predators, migrated northwards through western Canada in spring in search of fresh food supplies. Modern caribou do just the same today.

Finally, the fossil record shows that dinosaurs were only some of the victims of the dramatic events 65 million years ago. Other animal groups were drastically

(*Previous spread*) PRESERVED IN ROCK. The skeleton of an *Edmontosaurus*, one of the last dinosaurs, lies half-buried in rock exactly as it died.

reduced or disappeared. In the seas, all the plesiosaurs and mosasaurs vanished and the only group of marine reptiles to survive were turtles. Ammonites and belemnites (relatives of octopus and squid) were wiped out completely, together with most chalky plankton, brachiopods and clams. In the air, the pterosaurs disappeared but the birds did not. On land the mammals survived as did all other reptile groups (crocodiles, lizards, snakes, tortoises and turtles) as well as amphibians, insects and other invertebrates. Extinctions varied among major groups of plants, being most drastic among flowering plants, less so among conifers and least among ferns and mosses.

GOING EXTINCT. Rapid environmental changes can push species to extinction as conditions alter faster than the animals can adapt to meet them. *Bottom left:* The Pilori muskrat lived on the island of Martinique. The whole species was wiped out in 1902 when the eruption of Mount Pelée destroyed its habitat. *Bottom right:* The Irish giant deer was once found across Europe. The last survivors in Ireland died when a mini ice age destroyed their food supply. As with the dodo on Mauritius *(top)*, exterminated by sailors hunting it for food, human beings are responsible for most of today's extinctions, polluting and destroying natural habitats. The environmental catastrophe that overtook the dinosaurs was on a more global scale.

DEATH BY ISOLATION? In North America wetter and colder conditions may well have broken up the dinosaurs' habitat with swamps and rivers. Isolated plant-eating herds would have become cut off and died out one by one.

So much for the facts. Beyond this point everything is based on evidence that is incomplete and impossible to confirm. It is rather like investigating a murder committed 65 million years ago with no surviving witnesses and most of the scene of the crime under several layers of rock. The rest of this chapter takes a look at some of the most likely theories.

INTERPRETING THE EVIDENCE

As mentioned earlier, there is only one area in the world where the end of the dinosaur era can be extensively studied — the western interior of North America where the late Cretaceous and early Tertiary fossil record is most complete and the rocks are best exposed. New collecting techniques in this area have led to some surprising interpretations of the dinosaurs' disappearance.

In the past, palaeontologists had always looked for new or relatively complete fossil skeletons to excavate. Many other smaller pieces of fossil bone on the same site were ignored or left in the ground. The emphasis was on the quality of the specimens rather than the quantity. Yet these fossil 'leftovers' reveal important information as well. In particular, they can form the basis for estimating the number and type of dinosaurs that inhabited an area at one time. Collecting and identifying these fossil bones is rather like taking a population census — there is a margin of error in the data, but overall patterns and changes through time can be seen.

Over the last few years, American palaeontologists have undertaken systematic fossil quantity analyses of rocks from the very end of the Cretaceous Period in the Hell Creek area of northwest Montana, USA, using measurements like the number of teeth per tonne of rock or the number of different species per cubic metre. The results show that, in this area at least, dinosaur diversity (the number of different species) was in decline at the end of the Cretaceous, while mammals were increasing in number, variety and size to fill the gap. In other words, the dinosaurs were dying out for the whole of their last 5 million years or so. However, the last dinosaurs do seem to have met a fairly sudden end — sudden, that is, in geological terms. The rocks cannot tell us if the final extinction happened over weeks, months, years, or hundreds of years.

One reason could be that the climate changes during this period may have broken up the dinosaurs' usual feeding grounds into smaller isolated areas. Conditions generally became wetter with swamps and rivers fragmenting what had once been huge ranges. The large herds of plant-eaters could have found it increasingly difficult to survive, cut off from each other and having to compete for smaller areas of territory. Since plant-eaters were the base of the dinosaur food chain, their decline would have directly affected all the hunters and scavengers as well. In this hypothesis we see the dinosaur communities becoming isolated in small pockets of land and dying out one by one. Mammals began to take over the vacant space, but did not develop to their later levels of variety and number until some time after the last dinosaur had gone.

The dinosaur extinction comes at a point in the Earth's history when the Cretaceous rocks were being overlaid by the distinctive Tertiary rocks. This point is called the K–T boundary (K from the Greek word for chalk, *kreta*, after the great thickness of chalk rocks laid down in the sea by chalk plankton during the

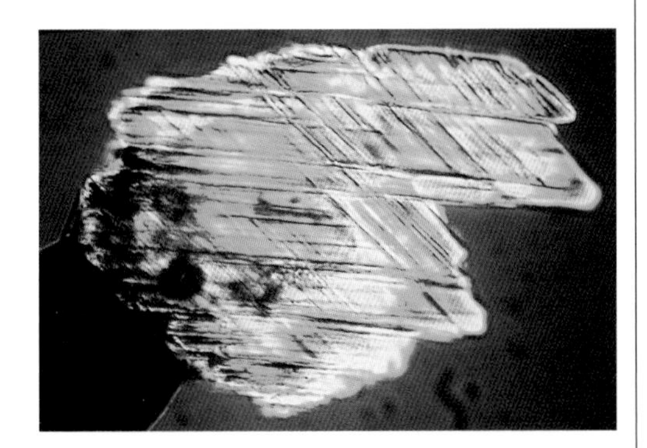

BLASTED ROCK. Distinctive parallel lines running across this 'shocked' quartz crystal show it has been formed during violent conditions of enormous temperature and pressure, such as a meteorite impact or a nuclear explosion. Similar crystals are found in K–T boundary rocks.

Cretaceous). It has been close examination of K–T boundary rocks around the world that has given rise to many new ideas about the mass extinction that included the dinosaurs.

THE ASTEROID THEORY

In 1978 Luis and Walter Alvarez, with a team from the University of California at Berkeley, USA, were studying the K–T boundary rocks at Gubbio in Italy. A layer of red clay about 2 centimetres thick just at the boundary point attracted their attention. Tests showed that it contained a level of the element iridium over 30 times higher than the average. Iridium is a very rare element in the Earth's crust. It usually arrives in the cosmic dust from space that is constantly showering the planet or, more unusually, from the Earth's core when certain types of volcano erupt. The high level 'iridium spike' led them to suggest that a huge asteroid, up to 10 kilometres across, must have hit the Earth 65 million years ago, leaving its iridium 'signature' in the rocks of the K–T boundary.

Such an event would certainly have been devastating for the dinosaurs. As

ASTEROID SIGNATURE. A pencil thin layer of clay marks the boundary between Cretaceous and Tertiary rocks. Its high iridium content can only have come from one of two places — a massive volcanic eruption from the Earth's molten interior or an asteroid strike from space. Either would have meant disaster for the dinosaurs.

DEATH FROM OUTER SPACE? A fireball passing through the Earth's atmosphere leaves a trail of increasing brilliance and explodes just beyond the field of view, South Galactic Pole, September 1991. Such a meteorite hitting the Earth may have resulted in: a primary blast wave equal to thousands of nuclear bombs; a chain reaction of volcanic eruptions; dust and gases filling the atmosphere blocking out the sun for months and causing acid rain; tidal waves sweeping back and forth against the coasts.

FERN INVASION. K–T boundary fossils show that a short-term fern takeover followed the dinosaur extinction. Similar 'fern spikes' occur after volcanoes erupt today, and indicate a major change to the environment.

the asteroid struck at 100,000 km/h a primary blast wave would have destroyed everything within a 400–500 kilometre radius. The same impact would also have sent tidal waves crashing against all the continents and set up a chain reaction of volcanic eruptions. The asteroid itself would probably have vaporized, sending up a huge cloud of dust, gases and water vapour into the atmosphere. The dust would gradually spread and cause a global winter of darkness lasting up to three months. The asteroid would also cause heating of the atmosphere, leading to chemical reactions producing acid gases such as nitrous oxide that would have washed out as acid rain.

Under these conditions no large animal, and no dinosaur large or small, could have survived. The land creatures best able to live through such an environmental disaster would be small active scavengers, able to find their food from a range of sources — mammals and birds, in fact. In the sea the plankton would die and the food chain that supported the large marine reptiles would collapse. The disturbance caused to shallow continental shelves as seawater, low in oxygen, was churned up from the deep sea, or acid rain fell, could account for the disappearance of the shelled sea animals — from plankton to ammonites.

Since the sensational publication of these ideas in 1979 (quickly dubbed the 'Worst Weekend in the History of the World'), other investigations of the K–T boundary have revealed further information. Most importantly, the 'iridium spike' that the Alvarez' discovered at Gubbio has been found at over 50 other places around the world, confirming beyond doubt that some extraordinary event occurred at this time. Microscopic examination of the sediments has also shown the presence of tiny grains of glass (quartz) with criss-cross fractures running across them. 'Shocked quartz', as it is called, is also found at the sites of nuclear explosions and its presence at the K–T boundary suggests there could indeed have been an explosive impact.

The plant fossils at the K–T boundary in North America also show evidence of short-term dramatic change. Just above the boundary, that is just after the event, there is a big increase in the number of fern spores found in the sediments. This has been seen as a rapid recolonization by the wind-dispersed ferns of the devastated mid-continental regions. The trees and shrubs were soon back in force, however, although not necessarily the same ones that were most abundant at the time of the catastrophe. Rare species, and species that were previously found in quite different areas, were the ones that seem to finally dominate the post-dinosaur landscape. The eruption of Mount St Helens in May 1980 illustrated dramatically what happens when a volcano erupts today, severely cutting back the surrounding plants and allowing the hardy ferns to dominate. After a short while the other plants grow back from buried seeds and rootstock, or seeds transported in from other areas, to regain their former balance. Again the K–T boundary gives us enough information to speculate but not enough to draw firm conclusions. There were no detectable changes at the K–T boundary to the plants growing at high latitudes in the southern hemisphere and yet the dinosaurs vanished there too.

If an asteroid did hit the Earth, where did it land? Estimates put the crater size at 180 kilometres across, but when the Alvarez' theory was published no positive sign of it had been found, even using satellite photography. This led to suggestions that the impact point was on part of the seabed that has since disappeared, but in 1990 the remains of an impact crater were found on the seabed off the northwest tip of the

LANDING SITE? *Above:* This crater, 1.2 kilometres wide and 170 metres deep, in Arizona was made by a small meteor. A hole 180 kilometres wide would have been punched in the Earth's surface by the K–T boundary asteroid. *Below:* Recent research suggests the asteroid may have landed in the sea just off the tip of the Yucatan Peninsula.

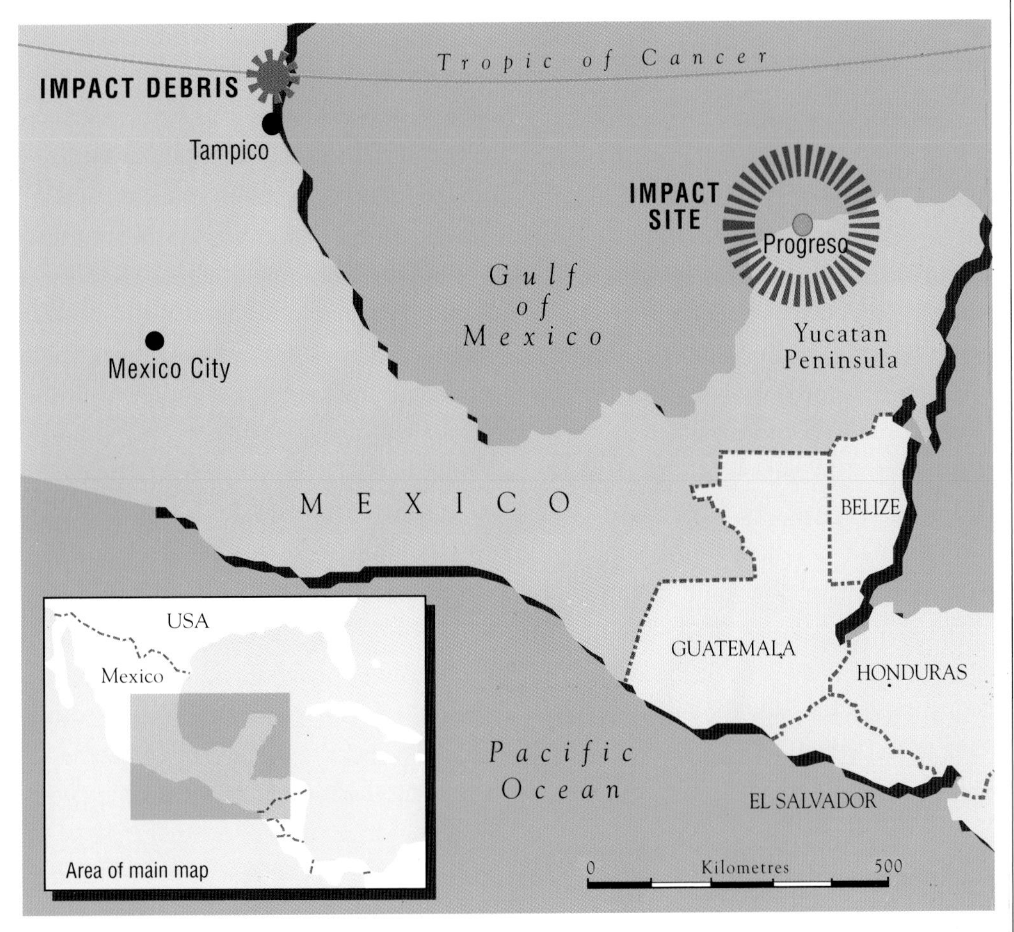

Yucatan Peninsula in Mexico. Possible debris from the explosion (shocked quartz and fossil tree remains mixed up with ocean bedrock) have been found up to 800 kilometres away, indicating the terrific power of the initial explosion.

THE VOLCANIC THEORY

Sampling of rocks above and below the K–T boundary have shown that the 'iridium spike' discovered by the Alvarez' was only one of several, separated by as much as half a million years. It is most unlikely that the Earth was struck by a succession of asteroids over this period, and so some scientists believe we should look to the other source of iridium for an explanation — the Earth's core.

These scientists argue that sustained volcanic eruption would have been enough to bring about the climate change that caused the dinosaur extinction. The effects on the environment of massive volcanic activity would have been quite similar to those of an asteroid impact. There certainly were huge volcanic eruptions in the Deccan region of India around 66 million years ago that went on discontinuously for a million years and produced flood basalt lava flows which now form the Deccan Traps — layers of rock up to 2400 metres thick. It is equally likely that these eruptions pushed the iridium rich lava to the surface and at the same time pumped huge amounts of carbon dioxide into the atmosphere, which ultimately led to acidification of the seas and the collapse of the marine ecosystem, and altered weather patterns. This pushed the dinosaurs beyond their capacity to adapt and survive.

DEATH FROM THE EARTH'S CORE? *Left:* Could volcanic activity 65 million years ago have altered the global environment, causing the death of the dinosaurs? *Above:* Deposits of cooled lava that form the Deccan Traps in India show that massive volcanic eruptions did occur around this time.

Volcanic activity may have affected dinosaurs in a quite different way as well — by destroying their eggs in the nest. Obviously the egg stage was the point in their life cycle at which dinosaurs were at their most vulnerable, a good reason for looking at their eggs as the possible 'weak link' that led to extinction.

The idea that eggs could be affected by volcanic activity is based on the fact that one of the rare elements released from deep inside the Earth by the Deccan Traps volcanism was selenium. Traces have been detected in K–T boundary sediments as far away as Denmark, and unexpectedly high levels were found in the eggshells from nests of sauropod dinosaurs from the latest Cretaceous in France. The amount of selenium in the shells increased in the eggs closest to the K–T boundary and these nests appear to have had a higher rate of hatching failure. Selenium is very poisonous, especially to growing embryos, and very small amounts in hen eggs causes them to die. Dr Hans Hansen, of the University of Copenhagen, Denmark, who carried out this work, has suggested that plant-eating dinosaurs suffered strongly reduced hatching as a result of eating large quantities of vegetation covered with volcanic dust containing selenium. This reproductive decline could in time have led to a complete break down of the food chain.

DEATH IN THE NEST

Another, now discredited, egg theory was that some small mammals developed the ability to crack open and eat dinosaur eggs, rather as a mongoose does with bird and reptile eggs today. Whole dinosaur generations would thus be wiped out before they even left the nest. Attractive though it might appear, this idea simply does not stand up to detailed study. It is possible that some mammals, and even some dinosaurs, were egg eaters. *Oviraptor*, for example had a powerful jaw that could have cracked eggs, and a *Velociraptor* skeleton has been found just above a nest of *Protoceratops* eggs. Yet modern egg eaters never actually destroy the species they feed off, since this would endanger their own food supply. It is not credible that the Cretaceous mammals would have disobeyed such a basic rule of scavenger survival.

So what really happened? No one knows for certain. There is clear evidence that the Earth's climate was changing towards the end of the Cretaceous Period. Also that two catastrophic events occurred — the sudden asteroid impact (in Mexico) and the longer term Deccan Traps volcanic activity. It may well be that the combined effects of all three events, in varying proportions in different parts of the world, proved to be too traumatic for the dinosaurs.

What we do know is that any theory which explains the disappearance of the dinosaurs alone is only telling us half the story. The extinction at the end of the Cretaceous shows puzzling patterns that neither a single big explosion nor a long drawn out decline can fully account for. What event, or combination of events, would prove equally fatal to a huge meat-eater like *Tyrannosaurus*, a medium sized plant-eater like *Edmontosaurus* and a fast running hunter like *Troodon*? What event would kill all dinosaurs and yet leave crocodiles unaffected — and, even more puzzling, allow the survival of amphibians? Frogs and salamanders are very sensitive to environmental changes and cannot breed in acid conditions. What event would kill flying reptiles like pterosaurs but not birds? What event would hit so many marine animals but miss sea turtles and reef corals? The only common pattern seems to be that it was fatal to be big. With very rare exceptions, anything over 1 metre long or 30 kilogrammes in weight on land, air or sea fell victim to the Cretaceous extinction.

As we undertake more detective work to unravel this 65 million year old extinction mystery it is worth remembering that, if it had not happened, we would not be here to investigate it at all. For it was the demise of the dinosaurs, and the huge gap that their departure left in the ecosystem, that allowed the mammals to evolve rapidly. They soon dominated the land and took to the sea and the air. Eventually, about 4 million years ago, the earliest ancestors of Man appeared.

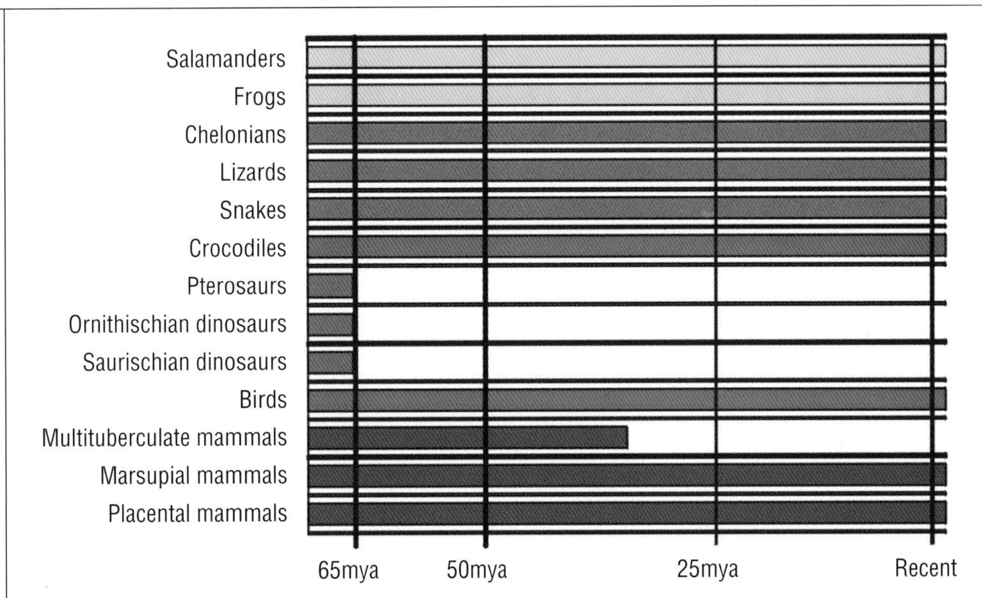

CUT-OFF POINT. The only groups of terrestrial vertebrates that did not survive beyond the end of the Cretaceous Period were three groups of archosaurs (red bars). Amphibians (yellow), reptiles (green) and two out of the three mammal groups (brown) are still living now.

THE MOMENT OF IMPACT. A scene such as this, resembling a vast nuclear explosion, may have happened as the K–T boundary meteorite landed in Mexico.

DINOSAURS A

8

The first dinosaur was named over 150 years ago. The subsequent history of dinosaur discovery has featured extra-ordinary people, unexpected finds and heroic achievements.

AND PEOPLE 8

Above all it is the story of how scientists have struggled to interpret scraps of evidence dug from the ground to understand the most amazing animals ever to live on Earth.

There were no human beings on Earth during the dinosaur era. Despite what you see at the cinema, the last dinosaur had been dead for at least 65 million years before the first *Homo sapiens* appeared on the scene. Yet in the short century and a half since dinosaurs were first discovered (or perhaps we should say recognized, since people had been discovering their fossilized remains for hundreds of years without knowing what they were), dinosaurs have fascinated scientists and the general public. Why? What is the attraction? Why are you, for example, reading this book? To answer these questions we first need to trace the steps by which dinosaurs came to be recognized, and how each step transformed people's understanding of the prehistoric world.

Dinosaur fossils have been weathering out of the rocks for millions of years. In ancient China they were thought to be the bones of dragons and were prized for their magic and medicinal powers. In fact, the very first description of a dinosaur fossil find comes from a book by the Chinese scholar Chang Qu, writing around AD 300, where he tells of the discovery of dragon bones at Wucheng in what is now Sichuan province. There are also records of American Indians discovering the bones of 'giant buffalo' which were thought to give the braves good luck during the hunt. These too were probably dinosaur fossils.

In England there are a number of old books and museum collections that contain pictures, descriptions and specimens of what we now know to be dinosaur fossils. The most famous is the description, by Robert Plot, of the end of a gigantic thigh bone from Oxfordshire which was first thought to belong to an elephant brought to England by the Romans, and later to be part of a giant human. His picture of the specimen strongly suggests that it was, in fact, from the meat-eating dinosaur *Megalosaurus*. In 1820 fossil bones found in the Connecticut Valley, USA were described as human remains (they actually belonged to the early dinosaur *Anchisaurus*). Between 1802 and 1860 literally thousands of dinosaur trackway fossils were collected in the same area, first by Plinny Moody and later by Professor Edward Hitchcock, under the impression they were the prints of giant birds. Indeed, Professor Hitchcock had a whole museum built at Amherst College to house his bird footprint collection and published books describing the prehistoric bird-filled landscape.

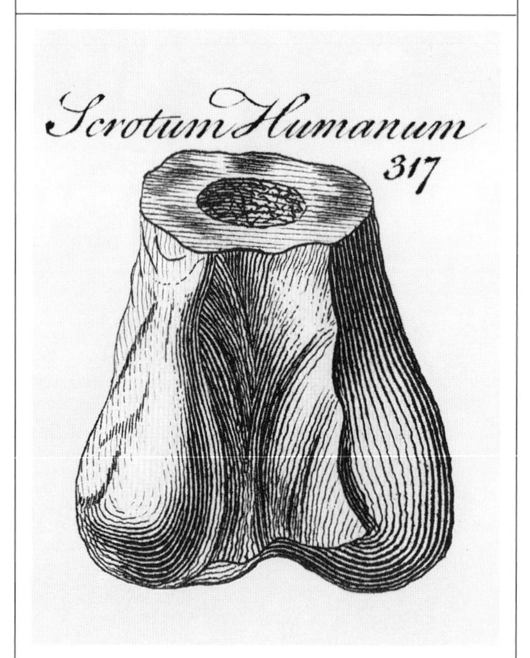

A GIANT HUMAN? Described by Robert Plot in 1676 as belonging to a giant human, this is the knee end of a thigh bone of *Megalosaurus*. It was later named, in 1763, by R. Brookes as *Scrotum humanum*, the genitals of a giant man.

With such evidence before them, why did scientists fail to recognize these fossils for what they were? Most people in Europe and America at this time accepted the traditional Christian teaching that the world was made by God in seven days and that all life started in the Garden of Eden. The seventeenth-century biblical scholar Archbishop Ussher had added up the ages of all the different prophets and patriarchs mentioned in the Bible, and concluded that the creation event took place in 4004 BC. The idea of extinct species of animals did not fit with these religious convictions. Why would God go to all the trouble of creating creatures and then let them die out?

Hard proof had been coming to light that the Earth's rocks were many

(*Previous spread*) A GIFT DINOSAUR. Andrew Carnegie presenting a plaster cast of *Diplodocus carnegii* to The Natural History Museum, London, in May 1905.

DRAGON BONES. *Left:* In ancient China dinosaur bones were thought to be the remains of dead dragons and were ground up for medicine and magic potions.

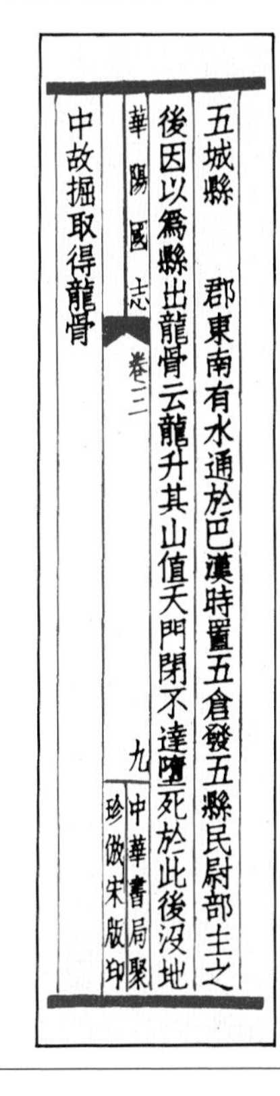

五城縣　郡東南有水通於巴漢時置五倉發五縣民尉部主之
後因以爲縣出龍骨云龍升其山值天門閉不達墮死於此後沒地
中故掘取得龍骨

華陽國志　卷三　九　中華書局聚　珍倣宋版印

THE FIRST RECORDED FOSSIL FIND. *Right:* Written over 1700 years ago, this account of a dragon bone find is the earliest written description of a dinosaur fossil.

millions of years older than the biblical date suggested and that some of these rocks contained the fossil remains of animals. This was controversial enough on its own, but the idea that prehistoric giant reptiles had inhabited the Earth millions of years before man was simply unacceptable. The work of the earliest dinosaur scientists thus not only shows brilliant deduction based on minimal fossil evidence, but also their intellectual courage in proposing extraordinary ideas about the Earth's prehistory that radically challenged the accepted thinking of the day.

GIDEON AND MARY MANTELL

The first person to recognize that huge reptiles had existed in the remote past was Gideon Mantell. Living in the town of Lewes, England, Mantell was a doctor with many patients in the surrounding countryside. He was also a keen amateur palaeontologist and often used his trips from one patient to another to scan the fossil rich Sussex countryside for specimens. The story is that, one day in the spring of 1822, Dr Mantell visited a patient with his wife Mary Ann. While he was in the house she strolled up and down the road, which was being repaired by workmen. A pile of stone from one of the local quarries had been dumped at the roadside and among the broken pieces she noticed something shining. She picked up the pieces of rock and saw that they contained some unusual fossil teeth. Mary Ann Mantell's decision to hold on to the specimens and show them to her husband was one of the most significant events in the whole history of dinosaur study.

Gideon Mantell instantly recognized that these were most remarkable fossils. He could tell that the teeth were from a plant-eater because they were blunt and had been worn down on one side by constant chewing. They must also have come from a very large animal indeed and looked, he thought, very like the incisor of an elephant or rhinoceros. With the help of the men who had delivered the stone, he tracked the source of the fossils to a quarry in the Tilgate Forest. Here the mystery of the teeth deepened, because the age of the rocks in the quarry, about 130 million years, showed that the teeth came from the Mesozoic Era — the age of reptiles when no large mammals (and certainly no elephants or rhinoceros) lived on Earth. Was it possible that these teeth belonged to a giant prehistoric reptile, hundreds of times bigger than any other so far discovered? We can imagine Mantell wrestling with this extraordinary idea, alone in his study in Lewes.

Mantell was convinced that his find was important but how could he find out more? He sent the teeth and some fossil foot bones found in the same quarry to two of the most famous scientists of the day, Baron Georges Cuvier in Paris and Dr William Buckland, the Professor of Geology at Oxford University. Their response was hardly encouraging. Cuvier wrote back to say that the teeth came from a rhinoceros and the foot bones from a hippopotamus. Buckland thought the teeth were from a large fish or much more recent mammal whose remains had been accidentally washed into the old rock of the quarry. Buckland even went so far as to warn Mantell, in a friendly way, not to publish a description of the bones.

Despite this reaction Mantell persisted. He began searching through collections of ancient and modern skeletons for anything that looked like the teeth and might give him an idea about the animal they belonged to. At the Hunterian Museum of the Royal College of Surgeons in London he was directed towards the vital clue, the skeleton of a modern iguana from South America, by a visiting anatomist Samuel Stutchbury. There, in its much smaller jaw, were virtual mirror images of the fossil teeth he had discovered. The same broad diamond shape, the same ridged edges. His prehistoric teeth were indeed from a reptile, some form of giant lizard, possibly like a huge iguana. In 1825, three years after his wife's first discovery, Mantell published a description of the find and the name of the animal, *Iguanodon* ('iguana tooth').

Mantell had to wait another nine years before making any significant progress on *Iguanodon*. In 1834 a large number of *Iguanodon* bones were discovered embedded in a slab of rock in a Maidstone quarry. The quarry owner demanded the high price of £25.00 for

GIDEON MANTELL. The first person to recognize the existence of extinct giant reptiles. These *Iguanodon* teeth puzzled Mantell for three years until he saw the jaw of a modern iguana lizard and made the connection between its teeth and those of the plant-eating *Iguanodon*.

the specimen and eventually some of Mantell's friends clubbed together to buy it for him. Now at last he could get a real picture of the size and shape of the animal he had named. His early sketches, based on the Maidstone discovery, show a crouching, lizard-like creature with a spike on its nose — hardly an accurate representation of *Iguanodon* as we know it today, but a huge leap in understanding at the time. The town of Maidstone in Kent, England, bears *Iguanodon* on its coat of arms as a tribute to the importance of this discovery.

BUCKLAND AND OWEN

In the study of dinosaurs (or any other organism) it is the date of publication of a specimen's description and name that counts as the moment of its official recognition. According to this rule, the first person to describe and name a dinosaur (even though he did not use the word) was not Mantell but Dr William Buckland, who published a description of the jaw of an animal that he called *Megalosaurus* in 1824, the year before Mantell's piece about *Iguanodon* appeared in print.

William Buckland was an eccentric clergyman (he later became Dean of Westminster Abbey) and geologist who had jackals in his living room and a bear in the hallway of his house. The bear also used to ride behind him on his horse and make frequent raids on the sweet counter of the local village post office. He was brilliant as well as eccentric and a well-known popular writer on scientific subjects ranging from the geological structure of the Alps to Ice Age mammals. His work on *Megalosaurus* was based on the study of

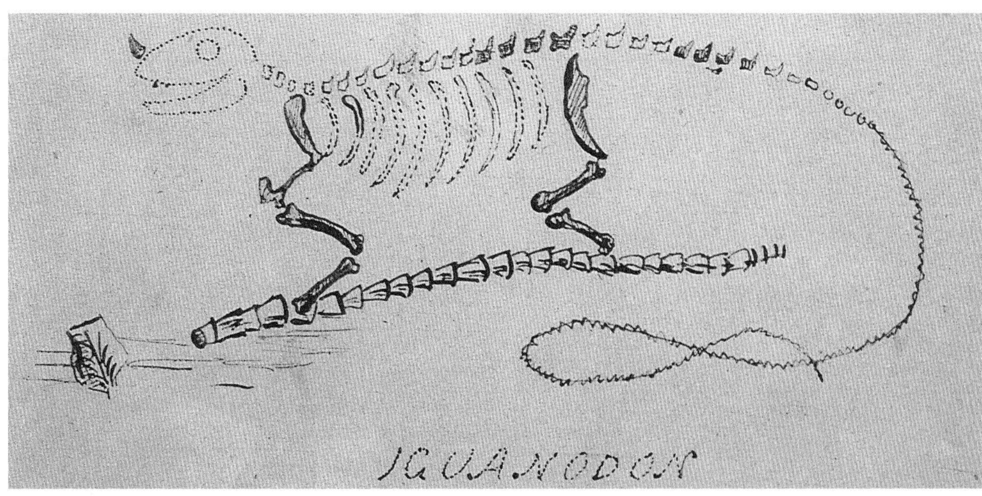

THE 'MANTELL-PIECE'. *Above:* "Now for three months hard work at night with my chisel", wrote the excited Gideon Mantell when this slab containing *Iguanodon* bones was delivered to his house in 1834 (spare a thought for his wife and children)!

Left: His early pen and ink sketch shows how lizard-like he thought *Iguanodon* to be. It was based on the only skeleton fragments then known and consequently was very inaccurate. For example, the thumb spike was mistaken for a nose horn and the ilium is drawn at the wrong angle.

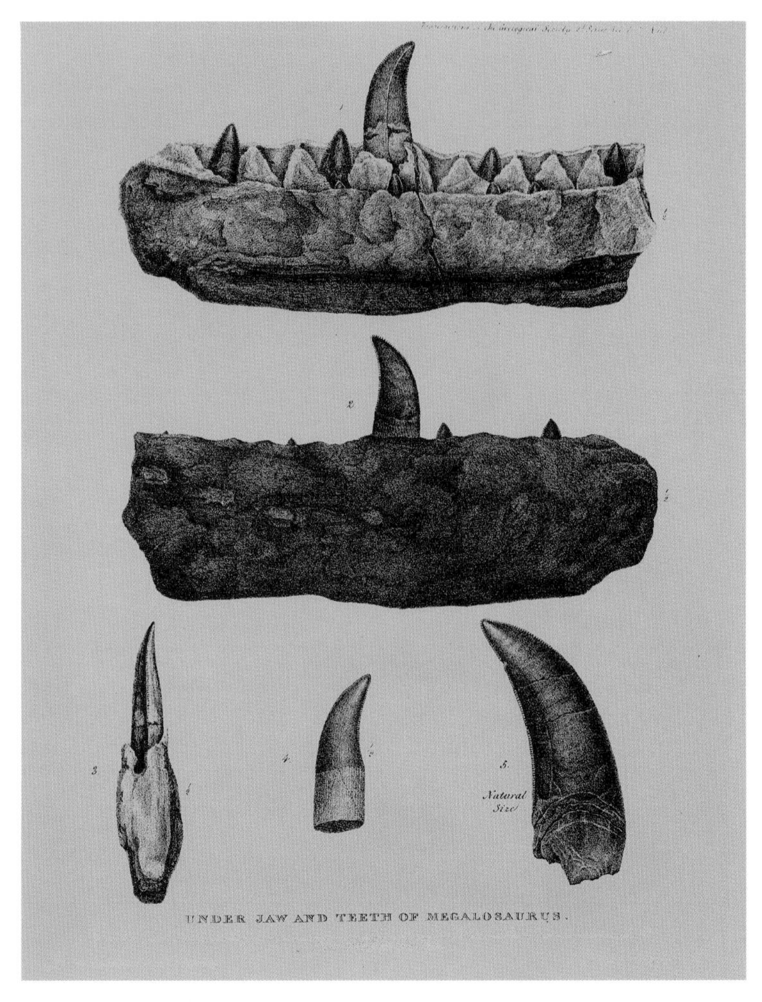

DEAN WILLIAM BUCKLAND. The first person to describe and name a dinosaur, *Megalosaurus*, which was based on remains including a lower jaw. A brilliant and eccentric man, he became Dean of Westminster Abbey and had his umbrella engraved with the words 'Stolen from Dean Buckland'.

the fossil jaw and bones that had been housed in the Oxford Museum since at least 1818. His published description of them is full and accurate, leading Baron Cuvier to estimate that *Megalosaurus* might have been up to 12 metres long, although he thought the remains belonged to a voracious marine reptile.

In the years following the publications of Mantell and Buckland, other giant reptile remains from the Mesozoic Era gradually came to light in England and Europe, and by 1840 nine such reptiles had been named. At this point the figure of Richard Owen enters the story. He was one of the foremost Victorian scientists and a world expert in the field of animal anatomy. He was a member of numerous committees and scientific organizations, a close friend of Queen Victoria (who presented him with a house in Richmond Park) and the first Superintendent of The Natural History Museum in London. One of his many appointments involved the dissection of animals that had died at London Zoo, so he had good first hand experience of unusual species and how their bodies were constructed.

Owen made an extensive study of all the giant reptile fossils available and came to the conclusion that they were not related to any modern reptiles. If they were not ancient versions of crocodiles or lizards, what were they? He reached the conclusion that three of them, *Megalosaurus*, *Iguanodon* and *Hylaeosaurus*, were prehistoric reptiles of enormous size belonging to a previously unrecognized group of animals he called Dinosauria from the Greek *deinos*, meaning 'terrible' and *sauros*, meaning 'lizard'. Owen first announced this name at the British Association for the Advancement of Science annual meeting in Plymouth, England, in 1841. It was a truly remarkable piece of deduction on Owen's part. Today, a scientist considering announcing a new

RICHARD OWEN. *Below:* Inventor of the word 'dinosaur' and the first person to recognize them as a distinct group. His designs for life-sized models, constructed by the sculptor Benjamin Waterhouse Hawkins, caused a public sensation when they were exhibited at Crystal Palace, London, in 1854. *Above:* This contemporary *Illustrated London News* picture shows a dinner in Owen's honour being given inside the half completed *Iguanodon.*

animal group would expect to study hundreds of specimens in microscopic detail to check and re-check the facts. Owen had the fragmentary remains of three quite different animals. He did not even really know what those animals looked like, as we can tell by looking at the famous life-sized restorations he supervised in 1854. These models, which can still be seen at Crystal Palace Park in London, show heavy, four-legged creatures with the build and stance of an elephant or rhinoceros. As we now know, these were very inaccurate, yet Owen managed to correctly identify dinosaurs as a single group and define several of the characteristics by which they are recognized to this day.

Over the last 150 years many brilliant people have found and named many dinosaurs. But only Gideon Mantell, William Buckland and Richard Owen can be said to have 'discovered' them. Owen's contribution was an intellectual one (he certainly never went digging for bones), building the idea of a new group of giant prehistoric reptiles from fragments of fossil bone. His brilliant thinking, Buckland's skilled observation, and Mantell's courage to press ahead despite discouragement, make these men unique in the history of dinosaur study.

FIRST FINDS IN THE USA

Soon after the elephantine dinosaur models were displayed at Crystal Palace, a spectacular dinosaur find in New Jersey, USA, cast doubt on the whole basis of Richard Owen's view of how dinosaurs looked. In 1858, William Parker Foulke was staying in the town of Haddonfield when he got into conversation with a local farmer. The farmer told how, 20 years earlier, he had dug up some huge fossil bones from a marl pit on his land and given them away to friends as souvenirs. Excited by the prospect of finding something unusual, Foulke got the farmer's permission to excavate the marl pit and hired a team of local diggers. They soon unearthed a large number of fossil bones and Foulke sent at once for the famous scientist Dr Joseph Leidy who was a Professor at the University of Philadelphia. Leidy studied the bones over the next few months and then published a description of them, naming the dinosaur *Hadrosaurus foulkii* (Foulke's heavy lizard). From the well-preserved leg bones Leidy was able to deduce that this dinosaur, which was in many ways similar to *Iguanodon*, stood upright in a bird-like posture, and was not a four-legged plodder, as Richard Owen had suggested. Once again, and not for the last time, the accepted view of dinosaurs was being challenged by the evidence of a new find. For the next 20 years this

LOUIS DOLLO. *Top:* Dollo spent most of his working life studying the *Iguanodon* skeletons found at Bernissart, Belgium. His work took dinosaur study to a new level of detail and accuracy. *Opposite:* The Bernissart skeletons were reconstructed in a church, the only large enough space, and were then transferred to the Brussels museum (*above*).

debate continued with each side finding and describing dinosaurs that seemed to prove their point. The fossil skeletons of the agile meat-eaters *Dryptosaurus* (formerly *Laelaps*) and *Compsognathus* were cited by one group as clear evidence of two-legged stance, while Owen's descriptions of the stegosaur *Omosaurus* (now called *Dacentrurus*) and the heavily armoured ankylosaur *Scelidosaurus* pointed in the other direction. Both groups were right, of course, as we now know from many finds of two- and four-legged dinosaurs.

THE BERNISSART FIND

The next major world dinosaur find was to resolve this and many other questions. In April 1878 some coal miners in Bernissart, Belgium, found themselves hacking through a giant fossil skeleton some 322 metres below ground level. They informed the mine managers who quickly handed over the supervision of the excavation of the bones to scientists from the Royal Museum of Natural History in Brussels.

What the miners had discovered was no less than a 'mass grave' of *Iguanodon* with nearly thirty complete skeletons, many of them preserved intact in the positions in which they died. More remarkable still, the fossils were not preserved in horizontal strata through the rock but in a vertical column of clay over 30 metres high that cut through the layers of shale and coal. How had this come about? Careful work over the next three years led scientists to believe that this had been a deep, narrow ravine in the prehistoric landscape. It was initially suggested that a herd of *Iguanodon* had fallen into it and been covered by layers of mud swept down by heavy rains. However, since the skeletons were found at different levels in the pit, it is now believed that the accumulation took place over a period of many years.

Piece by piece, block by numbered block, hundreds of bones were dug out of the mine and transported back to Brussels for study and assembly by the newly appointed 'assistant naturalist' Louis Dollo. This was an opportunity for the kind of dinosaur study that had never been possible before and very seldom since. Instead of a single animal Dollo had dozens of complete specimens to compare. He was able to prove conclusively that *Iguanodon* was two-legged, that the spike which Mantell and Owen had put on its nose was actually a thumb, and that the lattice-like ossified tendons along the spine were designed to give strength to the back and tail. He used the mass of fossil evidence to work out the position and size of muscles and the way that *Iguanodon* moved. At the same time he was busy reconstructing and mounting the Bernissart skeletons for display in the Brussels museum. About 30 of them can be seen there today, held in vast glass enclosures.

Louis Dollo's work on the Bernissart *Iguanodon* in many ways laid the foundation for the modern study of dinosaurs as living animals in a real environment. In addition to his study of the bones themselves he looked at the fossil remains of other animals buried with the *Iguanodon* skeletons, including several different species of crocodiles and turtles. He also studied the plant fossils and the way the sediments had been turned into rock, to try and construct as accurate a picture as possible of the dinosaurs and how they lived.

COPE AND MARSH

At the time of the Bernissart discoveries, dinosaur study in the USA was dominated by the personalities of two men — Edward Drinker Cope and Othniel Charles Marsh. Both were distinguished American academics, both brilliant, both hugely energetic in seeking and describing dinosaurs, and both hated each other. Their rivalry is one of the great legends of dinosaur science and was largely responsible for what has come to be called the 'American Dinosaur Rush'.

Why did Cope and Marsh detest each other? As late as 1868 there are records of them spending a friendly week together exploring the Haddonfield area (where *Hadrosaurus foulkii* had been found), yet two years later they were bitter enemies. One story is that Cope had described a new prehistoric sea

EDWARD DRINKER COPE

OTHNIEL CHARLES MARSH

COPE AND MARSH. Bitter rivals during their lifetimes, the names of these two great American palaeontologists will remain forever linked. Their rivalry certainly drove each of them to make extraordinary attempts to locate new dinosaur remains. Between them they named nearly 130 new dinosaurs, scouring North America for new species to put them ahead in the 'Dinosaur Rush'.

THE MILLIONAIRE'S DINOSAUR. Millionaire industrialist Andrew Carnegii financed an expedition to excavate spectacular dinosaur fossils in Wyoming, USA. One of them was named in his honour, *Diplodocus carnegii*, and he then paid for plaster casts to be made and sent to museums around the world. In London there was great public interest in the dinosaur when it was presented to the Trustees of The Natural History Museum (previously The British Museum (Natural History)) in May 1905.

reptile called *Elasmosaurus*, drawing particular attention to its strange backbone. Marsh came to see the fossil and pointed out that Cope had actually reconstructed it back to front with the skull at the end of its tail. Cope never forgave him for this humiliation. Whatever the reason, the rivalry between them was well established by 1877 when two schoolmasters, independently exploring the Colorado area, found large and exciting new dinosaur fossils. One sent his finds to Cope, the other to Marsh.

The result was an explosion of dinosaur exploration as Cope and Marsh hired teams of diggers to excavate the areas where the fossils had been found. Cope's patch seemed to have bigger and better bones, but later in 1877 Marsh had a stroke of luck that put him firmly ahead in the dinosaur race. In a mysterious letter, two men calling themselves 'Harlow and Edwards' told him that they had discovered giant bones at a secret location in Wyoming. 'Harlow and Edwards' turned out to be two railway workers. What they had actually discovered was a seven mile stretch of Jurassic fossil-bearing rock at Como Bluff alongside the Union Pacific Railroad. Marsh signed an agreement with them and they began shipping bones back to him by the tonne. Naturally, the Como Bluff area attracted Cope's attention as soon as he came to know what was going on there, and there are stories of clashes between the two rival teams as they searched for the best dinosaur remains.

By the end of the 1880s both Marsh and Cope had shifted their attentions to the Cretaceous fossil beds of western North America. Both of them had, in fact, made explorations here earlier in their careers. Marsh had named two dinosaurs from these rocks and Cope had undertaken an amazing expedition to Montana in 1876, the year when Chief Sitting Bull wiped out General Custer's forces at the battle of Little Big Horn. The Judith River fossil beds were in the heart of Sioux territory and the local people warned Cope against venturing out. He insisted it was safe, however, and in the end his only encounters with the Indians were friendly. One of his party describes how Cope used to entertain them by taking out his false teeth and showing them around. By 1892 both Cope and Marsh were back in the area again searching for specimens from the very last years of the dinosaur era.

Read any book on dinosaurs and the names of Cope and Marsh come up again and again. Their rivalry certainly spurred each of them into huge efforts to find new dinosaurs. Between them they named almost 130 new species, many from virtually complete fossil skeletons, transforming the understanding of dinosaur life. Louis Dollo had shown how dinosaurs could be studied as living animals based on the detailed study of one species. Cope and Marsh revealed the dinosaur world as a varied community of extraordinary reptiles. The contribution of these three men to dinosaur science has never been equalled.

DISCOVERIES ON EVERY CONTINENT

The first important dinosaur find in Canada was made in 1884 when the skull of the large meat-eater *Albertosaurus* was unearthed in the Red Deer River Valley in Alberta. Various attempts were made to explore this rugged terrain, but it was the American Barnum Brown who mounted the first full-scale expedition in 1910, using a huge, flat bottomed barge. This acted as a base camp as he floated down the river, stopping at likely points to search and excavate. The barge was also an ideal means of transporting the fossils themselves, which would certainly have been damaged by a long, bumpy horse and cart ride to the railway. Brown worked for several summers along the valley finding superbly preserved skeletons of dinosaurs such as *Corythosaurus*, *Styracosaurus* and *Centrosaurus*. Over many years the Red Deer River had worn a valley through the fossil bearing rock, exposing dinosaur bones at many levels along its steep banks. The valley rocks span millions of years of prehistory, allowing Brown to collect specimens that showed how dinosaur life in the area had evolved over time.

In 1907 a German engineer, working in what is now Tanzania, came across the fossil bones of a giant dinosaur just outside the village of Tendaguru. Word of the find soon reached the director of the Berlin museum who set about raising the money for a full-scale excavation. Money was certainly needed, for this was to be a dinosaur excavation of a scale and type never before attempted. Tendaguru was 64 kilometres from the coast — four days' march through jungle. There were no roads, no machines and no transport. At the height of the excavation over 500 local workers with their families were living on the site. It was a complete town that had to be provided with food, water, sanitation, building materials and

DINOSAURS GIANTS. No excavations on the scale of Tendaguru, Tanzania, had been undertaken before. Among the giant dinosaurs found there between 1909–1911 was this *Brachiosaurus*. Displayed at the Humboldt Museum in Berlin *(left)*, it remains the world's largest mounted skeleton. *Above:* This enormous shoulder blade was excavated by an expedition from London's Natural History Museum in the late 1920s.

medicines. The statistics compiled by the expedition organizers tell their own story about the effort involved. In the first three years (1909–1911) 4300 loads of fossil bones were sent from the excavation to the port of Lindi. Every load was carried by a team of porters who made a total of 5400 round trips. In the end over 250 tonnes of fossil and rock were carried away from Tendaguru. The finds were certainly extensive and spectacular. They included *Brachiosaurus*, the largest complete skeleton ever found, the plated stegosaur *Kentrosaurus* and the fast running two-legged plant-eater *Dryosaurus*.

The discovery of dinosaurs in Mongolia was almost an accident. In 1922, the American Museum of Natural History decided to mount an expedition to the Gobi Desert to search for early mammal fossils and the origins of prehistoric man. Roy Chapman Andrews was in charge of the practical arrangements and devised a plan whereby camel caravans plodded slowly through the desert carrying tonnes of petrol, oil, food and supplies. Meanwhile the fossil searchers covered hundreds of miles each day, zigzagging across the open terrain in cars and tracking their way back to prearranged meeting points using a nautical sextant and compass. The system worked beautifully but the scientists found very few human or mammal remains. Instead, in a series of expeditions between 1922 and 1925, they came across the skeletons of new and spectacular dinosaurs, including *Oviraptor* and *Saurornithoides*, and the first dinosaur nests and eggs, belonging to *Protoceratops*.

The American expeditions did a small amount of dinosaur exploration in China on their way to and from the Gobi Desert, and a Russian group excavated the Chinese hadrosaurid *Mandschurosaurus* (now called *Gilmoreosaurus*) in 1917. But in 1933 Professor Yang Zhong-jian was put in charge of Chinese dinosaur exploration

YANG ZHONG-JIAN. *Above:* CC Young (as he is better known in the West) was responsible for Chinese dinosaur exploration for forty years. *Below:* Finds of giant sauropods like *Mamenchisaurus* proved that related dinosaur species existed at the same time in quite different parts of the world.

ROY CHAPMAN ANDREWS. Seen here in the Gobi Desert with a nest of dinosaur eggs, Chapman Andrews was not himself a palaeontologist. His genius lay in organizing dinosaur hunting expeditions to some of the remotest parts of the world using cars, camels and porters.

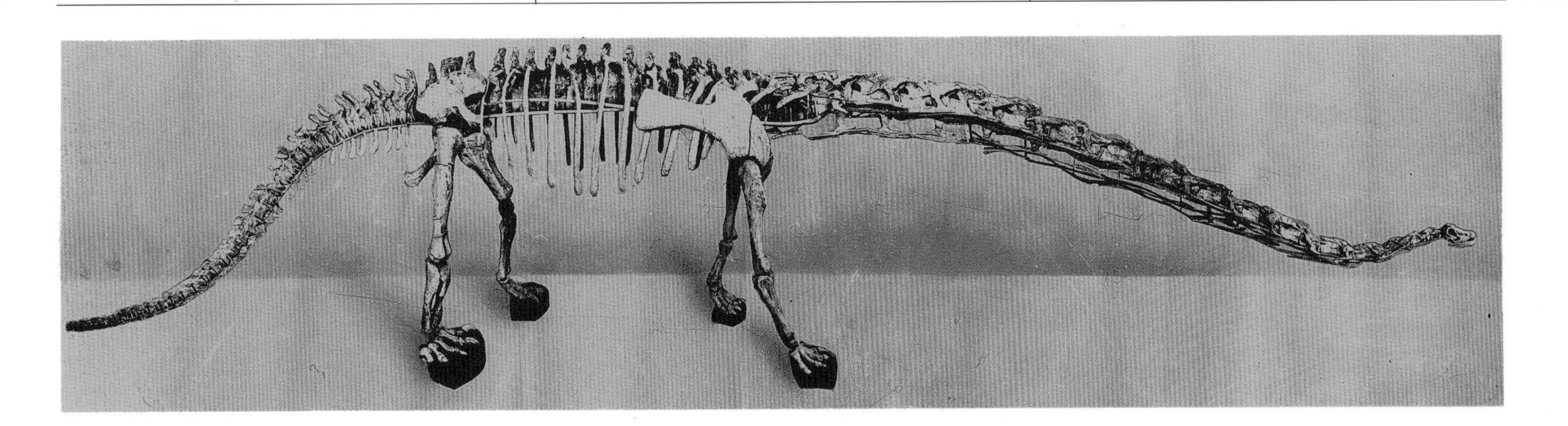

CHANGING VIEWS. This sequence of *Iguanodon* models shows clearly how scientific thinking about dinosaurs has developed over the years.

Waterhouse Hawkins, 1853, showed *Iguanodon* as a heavily built quadruped with its thumb spike mistaken for a nose horn.

Vernon Edwards, 1940s, basing his model on the Bernissart skeletons depicted *Iguanodon* in a strictly bipedal kangaroo-like posture.

Recent studies show that the weight of *Iguanodon*'s body was counterbalanced by the heavy tail and that it was able to move on two or four legs.

and, over the next forty years, was responsible for some of the most remarkable discoveries of this century. Dinosaurs from virtually every group have been found, including the amazingly long-necked *Mamenchisaurus*, the plated stegosaur *Tuojiangosaurus* and several hadrosaurs.

Today, work on dinosaurs goes on around the world with discoveries on every continent. Each new find expands what we know about the prehistoric world, sometimes confirming and sometimes overturning the accepted thinking about a particular species or aspect of dinosaur life.

Finds are made in the most far flung locations, such as the *Iguanodon* footprints discovered on the Arctic island of West Spitzbergen. At the other end of the world in Antarctica, early Jurassic dinosaurs were discovered in the Transantarctic Mountains in 1991. The locality, only 650 kilometres from the South Pole, produced dinosaur fossils belonging to three different theropods and a prosauropod. Most of the material was assigned to a new meat-eater, *Cryolophosaurus*, an allosaur-like animal with a furrowed bony crest above its eyes. The climate in Antarctica c.185 million years ago was quite suitable for dinosaurs.

Dinosaurs lived on the Antarctic Peninsula in late Cretaceous times about 70 million years ago. An ankylosaur and a small plant-eater similar to *Hypsilophodon* were discovered there in 1986 and 1989 respectively. At least one kind of hadrosaur lived there too — on the basis of a single but unmistakable tooth found in 1999. Antarctica was covered in cool temperate forests then, with summer temperature highs around 10° to 12°C and winter lows from 2° to 4°C. How did dinosaurs reach Antarctica? In the case of the hadrosaur, it could only have got there through a land bridge from South America, since hadrosaurs are found in North and South America and nowhere else on the southern continents. However, Australia and Antarctica were joined together in early Cretaceous times. The closest relatives of the *Hypsilophodon*-like animal lived in southern Australia, so the hypsilophodonts could disperse westwards into Antarctica.

Drs Pat and Tom Rich have spent years patiently excavating small dinosaur bones from impossibly hard rock on the sea coasts either side of Melbourne, south Australia. The fauna they recovered included several different small hypsilophodont plant-eaters, including *Leaellynasaura*. Part of the skull of *Leaellynasaura* was preserved complete with an endocast of the top of the brain (see Chapter 6 for more about dinosaur brains). The optic lobes, the brain's sight processing centre, were surprisingly large, which suggested that *Leaellynasaura* had enhanced visual acuity. These polar hypsilophodonts had correspondingly large orbits which must have housed large eyes. The conclusion is that they were adapted to living in dark or at least low light conditions. Southern Australia was sitting almost over the South Pole in early Cretaceous times and much of Australia lay within the latitude of 60° south, with a winter dark period of one to three months. The rocks in south Australia show evidence of episodes of freezing temperatures with average low temperatures of –2°C. In winter, the dinosaurs must have had to withstand cold conditions, and examination of their bone structure has shown that to be the case. The hypsilophodonts, and an ostrich dinosaur that lived there too, have bone typical of warm-blooded animals with a high metabolism, clearly an advantage in cold conditions. So at least some polar dinosaurs were specially adapted to high latitude living and perhaps even hibernated through the coldest period.

Migration would be one mechanism to avoid the worst of winter. Modern animals do that and there is some evidence that dinosaurs living in northern polar latitudes, as we saw in Chapter 7, did so too. Herds of plant-eaters may have spent the summer period in Alaska and moved south for the winter as caribou do today in North America. However, there is no evidence for either long-distance migration or year-round residents in southern high latitudes.

Amazing finds in South America have deprived *Tyrannosaurus* of its status as the largest meat-eater. *Giganotosaurus* now holds the record. It was first discovered in northwestern Patagonia, Argentina, in the early 1990s from rocks at the base of the late Cretaceous Period, between 112 and 90 million years old. Abundant remains of very large plant-eating sauropod dinosaurs were recovered from the same flood plain deposits. *Giganotosaurus* is estimated to have been up to 12.5 metres long with a weight of 6 to 8 tonnes — a little larger than *Tyrannosaurus*. The skull length, at around 180 centimetres, is the longest skull known for a meat-eating dinosaur, but it was lower than the relatively high domed *Tyrannosaurus* skull. The thigh bone (femur) of *Giganotosaurus* is 5 centimetres longer than the femur of the largest *Tyrannosaurus*, but the lower leg is shorter, making both animals about 4.6 metres tall at the hip. *Giganotosaurus* also has small arms but with three fingers instead of two, and its teeth are more flattened and dagger-like than the teeth of *Tyrannosaurus*. These and other differences between *Giganotosaurus* and *Tyrannosaurus* indicate that they were not closely related. Their lifestyles were quite different too. *Giganotosaurus*' teeth are quite short, narrow and uniform in size — adapted to slicing flesh. *Tyrannosaurus* had longer, broader teeth designed for a crushing bite through flesh and bone, as we saw in Chapter 3. *Giganotosaurus* must have been more of a meat slicer in keeping with its low, narrow, almost scissor-like skull shape. There was certainly plenty of meat to be had, up to 30 tonnes at a time! The giant *Argentinosaurus* that we met in Chapter 2, and several other sauropods, lived alongside *Giganotosaurus*. *Giganotosaurus* lived between 30 and 40 million years earlier than *Tyrannosaurus*. It is more closely related to the late Jurassic meat-eater *Allosaurus*, and represents a lineage of giant carnivores that evolved separately in the southern continent of Gondwana. *Giganotosaurus* has a very close relative in Africa, *Carcharodontosaurus*, that was only slightly smaller. Although known from fragmentary remains since the early 1900s, the first almost complete 160 centimetre-long skull was found in Morocco by American palaeontologist Paul Sereno in 1993. The discovery of these huge animals is quite fascinating; it shows how giant size was attained independently by meat-eaters in the two

SMALL PIECES. A British Antarctic Survey and Natural History Museum expedition collected a small plant-eating dinosaur on the Antarctic Peninsula in 1989. All the surrounding sediments were sieved to make sure that no tiny fragments of bone were left behind.

supercontinents of Laurasia and Gondwana. Despite their overall similarity because of their large size and the fact that both arose from bipedal ancestors, they evolved different ways of dealing with large prey.

Dromaeosaurs too have been discovered from South America. *Buitreraptor*, from the early Upper Cretaceous in Argentina, is the oldest known dromaeosaur from South America. It had a very long skull and small unserrated teeth in contrast to other dromaeosaurs. Its presence in South America provides evidence that dromaeosaurs originated further back in time before Pangea split into Gondwana and Laurasia. One branch of the family, the unenlagines, evolved in Gondwana in parallel with their counterparts in Laurasia.

The largest known dromaeosaur, *Utahraptor*, comes from the Barremian stage of the Lower Cretaceous Period in Utah, USA. No complete skeleton is known, but the remains include the characteristic huge curved claw on the second toe — up to 23 centimetres long. It also had extremely large, thin, blade-like hand claws. *Utahraptor* was 7 metres long and may have weighed up to 700 kg, making it a formidable predator. If it was a pack hunter like other dromaeosaurs, it has been speculated that packs might have been able to tackle a 20 metre-long sauropod, although on present evidence, it is impossible to know whether *Utahraptor* actually did live in packs.

A further clue to the brain power of dromaeosaurs comes from *Bambiraptor*, a diminutive sub-adult dromaeosaur from the Upper Cretaceous of Montana, USA that was less than a metre long and may have weighed about 2 kg. *Bambiraptor* had one of the largest-known dinosaur brains relative to its body size. The areas of the brain that deal with agility, co-ordination, intelligence and sight were enlarged and very bird-like. This suggests several life styles might have been possible – pack-hunting where co-operation and communication between individuals would have been vital as we saw in Chapter 6, or tree-climbing, as appears to have been the case in some other small dromaeosaurs described in Chapter 10.

Another recent South American find completely confounds the notion that sauropods all had long necks as described in Chapter 2. *Brachytrachelopan* from the Upper Jurassic of Patagonia, Argentina, completely reversed the trend towards neck elongation in sauropods. It was quite small by sauropod standards, around 9 metres long and was very short-necked. It appears to have been adapted for low browsing, feeding on plants growing at heights of between 1 and 2 metres. Oliver Rauhut has suggested that *Brachytrachelopan* might have had an ecological role equivalent to the low-browsing iguanodontians in the Upper Jurassic of Laurasia.

Dinosaur discoveries often seem geographically baffling until we remember that the patterns of land and sea were very different in the past as we saw in Chapter 1. The discovery of 70 million year-old late Cretaceous dinosaurs in Madagascar, now a huge island in the Indian Ocean, is still something of a puzzle to explain. A series of American-Malagasy expeditions, starting in 1993, found a huge diversity of animals including dinosaurs, crocodiles, birds and mammals. One of the most striking finds was a large, short-faced, meat-eating dinosaur called *Majungatholus* that belongs to a group of theropods called abelisaurs. Abelisaurs evolved in Gondwana and occur also in South America and India. *Majungatholus* is very like the advanced horned abelisaur *Carnotaurus* known from rocks of a very similar age in Argentina. However, India and Madagascar broke away from the rest of Gondwana about 140 million years ago, long before the time of *Carnotaurus*. India and Madagascar stayed as one continental block until India separated in the late Cretaceous. Madagascar has been isolated from the rest of the world's land masses for the last 140 million years. So how can the presence of *Majungatholus* there be explained? There are two possibilities, both involving the fragmenting of Gondwana. Either early abelisaurs were present all over Gondwana and separate parallel evolution took place in each land mass after they became separated by continental drift. However, if that was the case abelisaurs ought to occur in Africa — and Antarctica — which lay between South America and Madagascar. None have been found so far — although opportunities for fossil hunting in Antarctica are strictly limited! The second possibility is that *Carnotaurus*-like abelisaurs dispersed through the South America + Antarctica + Australia land mass in the early Cretaceous, after Africa had split away. They might have reached Madagascar (and India) by a temporary land bridge between Antarctica and Madagascar called the Kerguelan Plateau which occasionally surfaced at times of low sea level. Which hypothesis is most likely? Only the discovery of more fossils will provide the answer.

The therizinosaurs are really bizarre and little-understood dinosaurs, known mostly from the Cretaceous in Asia. *Therizinosaurus*, from the late Cretaceous of Mongolia, was up to 12 metres long, had a small head, a long neck, short tail and a large body. Its feet had very curved claws, and its most distinctive feature was gigantic hand claws, nearly a metre long. Some features of the skeletons of therizinosaurs are similar to those of theropods, yet they have a keratin covered toothless beak and small leaf-shaped cheek teeth — not something one would expect to find in a meat-eater. Only recently has it become clear where therizinosaurs fit within dinosaur evolution. *Beipiaosaurus* from Liaoning (see Chapter 10) although very fragmentary showed characters of maniraptorans (see cladogram on page 119), including shaggy feathers. Thanks to an unexpected find in Utah announced in 2005, the therizinosaur puzzle has become much clearer. *Falcarius* is the oldest-known member of the group from the early Cretaceous and the only record outside Asia. It was about 1 metre tall at the hips and 4 metres in length. Detailed analysis and comparison with other theropods place it as the earliest and most primitive therizinosaur and also clearly position the group within the Maniraptora. Moreover, the analysis shows that oviraptorosaurs and therizinosaurs are related. *Falcarius* demonstrates an early stage in the shifting of dietary habits from meat-eating to herbivory within the theropods. Some of the changes parallel those seen in clades of herbivorous ornithischian dinosaurs — keratin covered beaks, shredding teeth, changes to the pelvic girdle to accommodate a large gut, and toes adapted to increased weight support. So, not all theropods were meat-eaters after all! The nearest parallel for such a dietary switch among animals today is the Giant Panda from China which is a very specialised bear. Bears are

REVERSING THE TREND. Some Gondwanan sauropods shortened their necks and fed much closer to ground level in complete contrast to the long-necked *Diplodocus* from the northern continents.

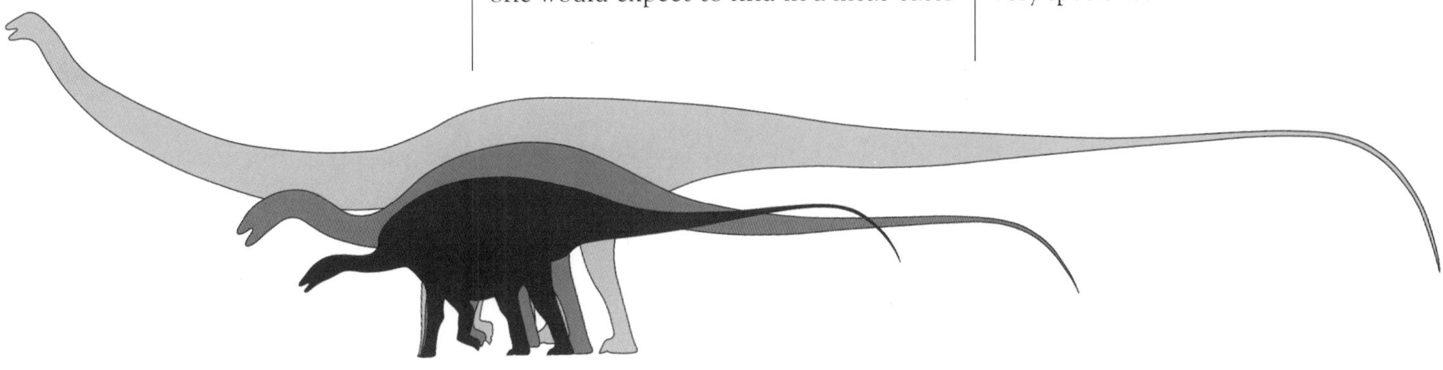

A DINOSAUR PUZZLE. *Therizinosaurus* is the most specialized member of a group of carnivores that switched to an herbivorous diet. It evolved several adaptations to plant-eating that parallel those in ornithischian dinosaurs, but no other dinosaur had such a spectacular set of hand claws.

omnivores, but the Giant Panda eats nothing but bamboo shoots and leaves.

So what did *Therizinosaurus* use its huge hand claws for? Perhaps for gathering and pushing vegetation into its mouth. The claws might also have been used for defence against predators, such as the contemporary *Tarbosaurus*, a close relative of *Tyranosaurus rex*, or even perhaps for territorial and mating battles.

New discoveries are being made world-wide every year and new techniques of laboratory study are yielding more information from fossils that have sat in museum storerooms for decades. New ideas are being published continually, new research undertaken and new theories discussed that take us step by step nearer to an understanding of the dinosaurs and their world.

AN ENDLESS ATTRACTION

Dinosaurs have always fascinated the general public. Only 14 years after Richard Owen first invented the word his life-sized restorations were attracting huge crowds into Crystal Palace. Today, a new dinosaur exhibition will guarantee queues around the block for any museum in the world. More than anything else people want to see what dinosaurs looked like, and the work of skilled dinosaur illustrators has always been in high demand. Some details about how illustrators work are given in Chapter 9, but it is very important to remember that each picture is only as accurate as the scientific information that the artist is given.

Alongside the popular interest in dinosaur science has come the extensive use of dinosaurs in fiction and adventure stories. Dinosaurs are a gift to authors and comic strip artists because they create an instant adventure, being big, dangerous and liable to fight anything that moves. They can also be portrayed as slow and stupid, allowing the human heroes to outwit them and triumph in the end. Scientific accuracy plays little part in these stories. Once the humans and dinosaurs have somehow been brought together, then dinosaurs from any and every era are likely to appear alongside each other displaying the kinds of behaviour that can make even the most broad-minded palaeontologist howl with anguish.

The appearance of dinosaurs in feature films and cartoons, often portraying them as comic and endearing, has added a whole new dimension to the task of illustration. The very first use of dinosaurs in moving cartoon form, 'Gertie the dinosaur' by Winsor McCay in 1914, used just this sort of technique to appeal to cinema audiences. A number of techniques to show dinosaurs moving were tried in the early days, many of them pioneered by the American special effects experts Willis O'Brian and later Ray Harryhausen. From moving models to trick optics and remote controlled robots, the quest for realistic screen dinosaurs gets ever more complex (and expensive).

While dinosaurs are an accepted fact of prehistoric life for most people there are those who find it hard to accept the scientific facts about their existence. Such people usually also reject the whole concept of evolution, insisting instead on religious beliefs about the creation of the whole world as a single, divinely organized event. This has led to some interesting interpretations of dinosaur fossil evidence. In one location in Texas, USA, a set of sauropod trackways is said by many 'creationists' to show the footprints of human beings walking beside the dinosaur. This is clear evidence, they claim, that dinosaurs and people did indeed live at the same time. Even today dinosaurs can be an issue of intense controversy and passionately held beliefs.

Which brings us back to the question that started this chapter. Why are people so interested in dinosaurs? Firstly, because dinosaurs were such extraordinary animals. They have something of the fairytale dragon or monster about them which is exciting, frightening and fascinating all at the same time. Secondly, they are all dead so we do not have to worry about what they might do to us or what we are doing to them – a depressing anxiety with the more extraordinary animals in today's world. Most importantly, dinosaurs are so popular because there is so much still to find out about them. Every new fossil find has the potential to challenge the accepted wisdom about a dinosaur species or even about dinosaurs in general. What if fossilized skin were found with the colour somehow preserved? What if a family of *Tyrannosaurus* hatchlings was discovered? Or evidence of even larger sauropods than *Argentinosaurus*? Anyone could make such a find. Amateur fossil collectors are the eyes and hands of dinosaur scientists around the world. No other branch of science offers such a potential opportunity for non-experts to participate or such exciting rewards for doing so.

PIECING T

9

What type of work do palaeontologists do? The journey from first finding a new dinosaur fossil to putting its skeleton on display can take many years. In between lie painstaking

OGETHER 9

hours of preparation, restoration, academic study and writing. The skills of many people are employed to create as detailed a picture as possible of the living animal.

Studying dinosaurs requires an extraordinary range of skills. A good palaeontologist must combine the qualities of an explorer, an archaeologist, a detective and an artist as well as having a detailed knowledge of animal structure and function. So what do these dinosaur scientists do? How do you study a dinosaur?

First you have to find one, or rather find its fossilized remains. Fossils are only formed in certain kinds of sedimentary rock (see Chapter 1), and there are only a few places in the world, some very remote, where that rock is exposed at the surface. The good news is that new dinosaur fossils are being exposed all the time as wind and water wear away the covering rock. Some dinosaur discoveries are made quite by chance when amateur collectors explore old and new fossil beds. Others are planned, usually when a museum or university mounts a full-scale expedition to a carefully chosen site, complete with trucks, generators, and tonnes of digging, surveying and packing materials.

FOSSIL FINDS. *Below:* Dinosaur discoveries have been made on every continent. Some of the fossil sites are very remote indeed such as northern Alaska and Antarctica.

THE PALAEONTOLOGIST AT WORK

The purpose of every expedition is to excavate new dinosaur skeletons and bring them back to the laboratory for study. Half a hillside of rock may have to be removed, yet the fossil bones themselves must be treated with the utmost care. The palaeontologist may use a pneumatic drill one day and a putty knife and soft paintbrush the next. In the final stages it is quite common to cut out the fossil, still embedded in rock, and pack it up in plaster of Paris bandages or polyurethane foam for transportation. It is vital, however, that a detailed survey is made before anything is removed from the site. The survey notes the exact position of all the bones when found, and can prove invaluable in later study of the dinosaur skeleton. The lumps of fossil rock are then taken away, usually in trucks, although camels, mules and elephants have also been used!

Back in the laboratory the palaeontologist–explorer turns into a palaeontologist–preparator, whose task is to remove the rest of the surrounding rock, called the matrix, from the fossil bones. These may have been broken or crushed during fossilization, so the matrix must be taken away very gently to avoid losing any fragments. The one thing in the palaeontologist's favour is that the matrix is often (but not always!) a softer rock than the fossil inside.

A typical lump of rock, fresh in from the excavation, will undergo a strict preparation process. Large pieces of the matrix are first removed with hammer and chisel, a diamond-bladed pneumatic saw or even a shot-blaster, leaving a thin coating of rock round the fossil. There is now a choice. If the matrix is made of limestone it can be dissolved away with a solution of acetic or formic acid, allowing the acid-resistant fossil bone to

(*Previous spread*) DEATH POSE. A scale model of the British dinosaur *Baryonyx* based on years of work by scientists at The Natural History Museum, London.

THE *BARYONYX* STORY: DISCOVERY. In January 1983 Mr William Walker, an amateur collector, found this huge fossil claw bone in a Surrey clay pit (shown here actual size). Palaeontologists at The Natural History Museum in London identified it as belonging to a large and unusual meat-eating dinosaur. No one yet suspected that Mr Walker had made the most important European dinosaur discovery of the century.

emerge gradually. This technique is particularly good for small, delicate fossils but it is very slow.

If the matrix is very thick or insoluble in acid, the shot blaster may well be used. This is housed in a huge steel and reinforced glass cabinet in which cast iron shot or plastic pellets about the size of sugar grains are fired at the fossil rock, with a noise like a violent hailstorm. Great skill is needed to manipulate the rock during blasting so that the fossil bones are exposed without damage. As each part of the fossil appears it is painted with a coat of latex rubber to prevent it being eaten away by the pellets. It is rarely possible to deal with the whole matrix by blasting, but it can rapidly remove most of the rock 'coat' and reduce the rest to a few millimetres thickness. Sometimes thick rock is removed by cutting grooves in it with an air-powered saw and then chipping away the ridges in between. This process can be repeated several times. In this way the rock is removed in controlled layers.

It is now that the really painstaking work begins. With a fine air-powered engraver, dentist's drill, scalpel and needle, the palaeontologist removes the last of the matrix (sometimes as little as one grain of rock at a time), strengthening the fossil with fast drying resin as the bone is uncovered. Remember, dinosaur bones are not conveniently smooth and straight. They have hollows, holes, curves, cracks and little sticking out bits that can drive you mad. A week's intensive work under the stereo microscope may do no more than expose the details of one vertebrae (*Diplodocus* had over ninety), and it can take years of labour to prepare a whole skeleton. Every dinosaur museum in the world has dozens of mysterious plaster of Paris packages on its store shelves — untouched fossils just waiting for preparation, 'when there's time to get round to it'.

SCIENTIFIC STUDY

By now the palaeontologist has changed roles again and is beginning the academic study of the fossil specimens. The first stage is to identify just what all the pieces are. It is very rare that a dinosaur skeleton is found complete with all the bones helpfully in the right place. Pieces of the dead animal may have been carried off by scavengers, washed away by a river, re-arranged or destroyed as the surrounding rock shifted over millions of years. The result is a three-dimensional jigsaw puzzle which may have any number of pieces missing or broken, and from which a reconstruction is attempted without a guiding picture.

THE *BARYONYX* STORY: EXCAVATION. *Above:* It took The Natural History Museum team three weeks to excavate the skeleton. Some fossil bones were loose in the clay, but most were encased in hard blocks of iron-impregnated siltstone. Each was separated from its neighbour and packed in a protective jacket for transportation. *Right:* Before leaving the site the exact position of every block and bone was meticulously noted by the palaeontologists on this survey map.

METRES

- ■ skull
- □ lower jaw
- ☆ loose teeth
- ▲ cervical vertebrae
- △ dorsal vertebrae
- ▼ caudal vertebrae
- ▽ ribs
- ▶ abdominal ribs
- ◆ pectoral girdle
- ● forelimb
- ◇ pelvic girdle
- ○ hind limb

Just like doing a jigsaw, the palaeontologist starts with the important pieces — the ones that will give most information straight away. Skulls and teeth rank first among these so called 'diagnostic' bones. Each main group of dinosaurs had a very different skull structure, and it is often possible to find out a lot about your new discovery from quite small pieces of skull. Teeth give you information about the dinosaur's diet, and vertebrae are also useful, again because each dinosaur group had its own particular kind.

The palaeontologist's work at this stage is very slow and careful. Each piece of bone must be described and illustrated in detail. Part of the description of a single bone might read like this:

". . . the long strongly opisthocoelous cervical vertebrae with short neural spines, with well-developed epipophyses, and with ends of the centra not 'offset' so that when they are correctly apposed . . . "

This is pretty heavy going, particularly when you consider that a large skeleton may have more than 300 individual bones to describe.

The technique for illustrating bones has not changed much since Victorian times, with pen and ink being most often used. Some palaeontologists are superb scientific artists as well and do all their own drawing. Others work with skilled illustrators to give a view of all the bones from several angles. A detailed photographic record, sometimes stereoscopic, is made at the same time.

The process of identifying and describing the fossil bones often throws up as many questions as answers for the palaeontologist. It is now time to tackle these questions using a mixture of scientific principles, deduction, comparison with other fossils and with modern animals and sometimes, inspired guesswork. For example, how do all the pieces of the bone jigsaw actually fit together? If the palaeontologist is lucky there may be up to half the complete

THE *BARYONYX* STORY: PREPARATION. A shot-blaster eats away at a block of rock. Most of the siltstone matrix surrounding the fossil bones was removed in this way, with power chisels and diamond-bladed saws as backup. Some blocks contained several bones or parts of bones. The exposed fossil bone was coated with latex rubber to prevent damage during shot-blasting.

THE *BARYONYX* STORY: MORE PREPARATION. Working under a stereo microscope, a scientist scrapes away the last traces of matrix with an air-powered engraver. Quick drying resins were used to strengthen the fragile bone as it appeared. Every detail of the new fossil was precious because of the information it could give about the dinosaur. The fully prepared pieces were placed in air-conditioned storage to await detailed study.

THE *BARYONYX* STORY: RECONSTRUCTION. It is often necessary to model pieces of missing skeleton or make casts of the fossil bones for detailed study. Modelling is done with softened wax or modelling clay, gradually adding the missing pieces to the original. Moulds are made from silicone rubber and may consist of several interlocked pieces. Bone-coloured resin and glass fibre are used for the casts, which are exact replicas of the originals.

skeleton to work with. Quite a few gaps can be filled by modelling from the bones that are present. A right rib, for example, can be carefully modelled as a 'mirror image' of the corresponding left rib. If you have three toes from one foot and one from the other you can use them as the basis for making up a complete set for both. But if a great number of bones are lacking, you have to fill them in by scientific comparison with the most similar dinosaurs and by using the site survey map to see how the skeleton lay when it was found.

You can see now how important the site survey is. Careless collecting and surveying can lead to some amazing mistakes at the reconstruction stage. Bones from two different dinosaurs found in the same place can be assembled as one, or pieces fitted into the wrong place entirely. Indeed, until very recently the original skeleton of *Apatosaurus* (discovered by Othniel Marsh in 1877) was restored with a *Camarasaurus* skull. This was because an isolated skull and headless skeleton had been found quite close together and it was assumed, following Marsh's notes,

THE *BARYONYX* STORY: DESCRIPTION AND ASSEMBLY. The white areas are the parts of the dinosaur found during the excavation, about 60 per cent of the whole skeleton. By carefully examining and describing each fossil bone, palaeontologists were able to fill in the missing parts and assemble a picture of the complete animal. The dinosaur weighed about 2 tonnes, was over 10 metres long, 3–4 metres tall and it walked on two legs.

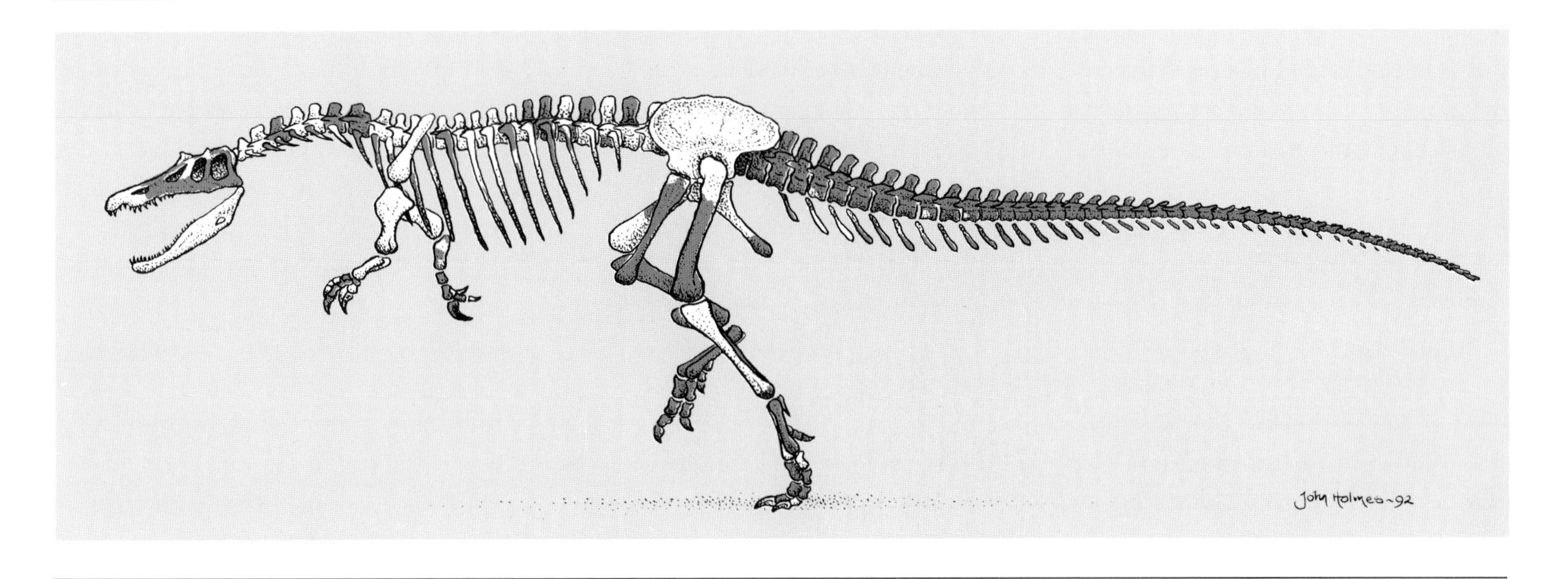

that they belonged to each other. It was not until 1975 that this mistake was finally corrected and *Apatosaurus* acquired its proper skull.

If the dinosaur you have discovered is one of an already known species you have the work of other scientists to guide you. This can be no more than a guide, however, since your example may have new pieces or even shed new light about the whole way the animal stood or walked. You need only look at the way our view of *Iguanodon* has changed over the years to see that there is always room for new thinking, even with the best known dinosaurs. If your discovery is a new species, reconstruction is more challenging still and you must use what you know of other similar dinosaurs as a basis.

With the skeleton assembled as a working drawing, the palaeontologist now tries to put the muscle structure in place. There are no such things as fossilized muscles, of course, and even a simple action, like chewing, involves a large number of muscles working together. Comparison with living animals is the only real way to understand this complex aspect of the dinosaur, and a detailed knowledge of modern animal anatomy is essential for every palaeontologist.

Finally, the three-dimensional jigsaw puzzle is completed. Each piece of fossil bone is identified and missing pieces filled in where possible. The basic size, shape and stance of the dinosaur are clear. It may have taken the palaeontologist years to get this far, but it is now possible to attempt some further detective work, drawing on all the available evidence to deduce more about how the dinosaur looked and the way it might have lived.

THE *BARYONYX* STORY: INTERPRETATION. *Top:* The s-shaped curve of the new dinosaur's jaw was strikingly similar to that of modern fish-eating crocodiles. It also had 64 teeth in its lower jaw, twice the number found in other meat-eating dinosaurs. Palaeontologists deduced from these features that it must have been a fish-eater. *Bottom:* This idea was later confirmed when fossil scales and teeth of the fish *Lepidotes,* like the one shown here, were found in the area where the stomach would have been.

BUILDING UP THE PICTURE

As with any detective, the best place to search for clues is where the body was found. Skilled interpretation of the surrounding rocks can sometimes show the position of cliffs, valleys and waterways, and whether the dead dinosaur was buried in sand, mud or river silt. This gives an indication of what the prehistoric location might have looked like, and even what the dinosaur was doing there. Sometimes the evidence from the rocks is very dramatic. At one site in Utah, USA, a large number of different sized *Allosaurus* skeletons have been found together (the only mass discovery of large meat-eaters) as well as the remains of different plant-eaters including *Stegosaurus* and *Camarasaurus*. The surrounding rocks show that the area may well have been a natural bog, so presumably one *Allosaurus* was attracted by a stranded plant-eater and became trapped in the sticky ground. Perhaps one by one the other hunters waded in to get a bite of the plant-eater

SURREY SCENE. Life in southern England 124 million years ago. *Baryonyx* fishes on a river bank fringed with horsetails and ferns as a crocodile swims by, while brachiosaurs browse on the evergreen trees, *Polacanthus* walks up the river bank and pterosaurs fly overhead.

and each other and became stuck themselves, finally suffocating or starving to death. It is possible that this place was a natural hazard for many years, accumulating its remarkable skeleton collection over time.

Detailed study of individual skeletons can lead to other questions about the history of all dinosaurs. An example of this is the skulls and teeth of the two main groups of plant-eaters. In the Jurassic Period sauropods such as *Diplodocus*, with their simple raking or chopping teeth and fermentation digestion, were the most abundant plant-eaters. Their decline at the end of the Jurassic is mirrored, in the early Cretaceous, by an increase in the abundance of ornithopods like *Iguanodon*, which had grinding teeth and processed the food in their mouths. Flowering plants also made their appearance at this time. Is it possible that the grinders were better able to cope with the new kinds of plant food? This seems distinctly possible, especially as the most efficient grinders of all, hadrosaurs like *Edmontosaurus*, became the biggest and most successful group of plant-eaters in the late Cretaceous. Perhaps the comparison of skulls and teeth actually tells this much wider story of dinosaur evolution.

The other key task that the palaeontologist must undertake is to work out where an individual species should be placed on the dinosaur 'family tree'. Many learned and weighty volumes have been written on this subject, but there are basic principles to help sort out the main relationships.

What we need to look for are characteristics that made one kind of dinosaur different from all the others, and characteristics that several species had in common. These characteristics must ideally be really major ones, fundamental to the way the dinosaur is constructed, and not just incidental features. And they must be to do with bones and teeth, since that is all we have left of most dinosaurs. Dinosaur sorting or classifying, therefore, involves looking at each dinosaur and asking a series of 'Does it have?' questions about its fossilized skeleton. The more questions you can answer, the better you can narrow down its position on the family tree.

The first question is, 'Does it have hips like a bird or like a lizard?' Bird-hipped dinosaurs (which are not related to birds despite the name) belong to the group called Ornithischia and are all plant-eaters. Lizard-hipped dinosaurs belong to Saurischia, a group that contains both plant- and meat-eaters. The diagram shows the kind of further questions that must be asked about the dinosaur, all of them related to teeth, skulls, vertebrae, claws and so on. Some of them may be

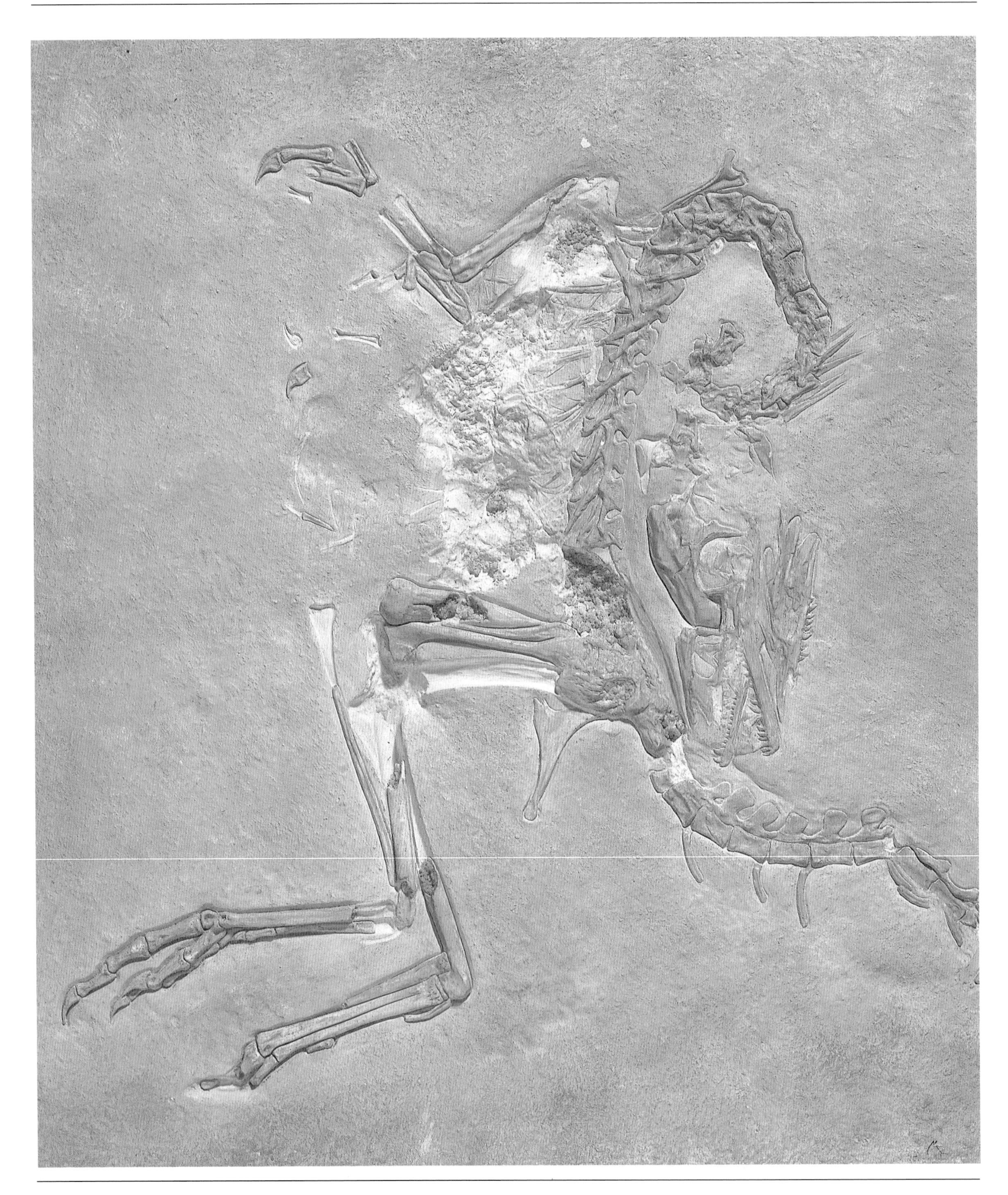

NAME	MEANING
Tyrannosaurus rex	King tyrant lizard
Diplodocus carnegii	Carnegie's double beam
Deinonychus antirrhopus	Counterbalancing terrible claw
Iguanodon atherfieldensis	Iguana tooth from Atherfield
Iguanodon bernissartensis	Iguana tooth from Bernissart

DINOSAUR NAMES. Dinosaur names have two parts, both derived from Greek or Latin, as do all animals and plants. The first name identifies the genus and the second the species. In many cases there is only one dinosaur species in each genus so just the first name is used as a 'short-hand' version. When palaeontologists choose the name of a new dinosaur they try to make it say something significant about the nature of the animal, where it was found or who made the discovery.

very difficult to answer indeed, especially if there are vital bones missing from the specimen. But it is only on this basis that the dinosaur can be accurately classified.

The palaeontologist is finally sure which main group the dinosaur belongs to. Now the skeleton must be minutely checked to see if it is one of a known species within that group or something that no one has described before. Again, this may be a very tricky task. After all, just when is a dinosaur different enough to count as 'unknown' rather than just a variation of 'known'. Sometimes even named specimens are re-examined and identified as being something else. There have been heated debates and even life-long disputes between palaeontologists on this subject.

Naming and publication of the dinosaur's description are the last duties of the palaeontologist. This is a very formal process in which the bone by bone description must follow a strict formula with carefully reasoned deductions about the dinosaur itself. A panel of other experts reads and approves the paper before it is published in a scientific journal. The existence of the new dinosaur is recognized from the exact date of publication. This may seem to involve a great deal of extra work but it is the palaeontologist's most important task because it allows the new discovery to be understood by other scientists. With such a very small number of dinosaur skeletons to study (less than a dozen specimens of *Tyrannosaurus* have been discovered in the last 150 years, for example, and only two or three reasonably complete skeletons), it is vital that palaeontologists should know about what has been found in other parts of the world.

BRINGING DINOSAURS TO LIFE

Modelling the dinosaur and putting the skeleton on public display require yet more skills. Most full skeletons that you see in museums are actually replicas of originals. Fossil bones can indeed be mounted on heavy metal frames called armatures, but their weight and fragility make them difficult to display safely. Replicas, which can be made of light, hollow materials, such as glass fibre, and which weigh a fraction of the original, can be held in place with much thinner metal rods or plastic tubes, often running invisibly through the inside of the bones. The palaeontologist will work with a team of technicians to ensure that the dinosaur is displayed safely, securely and in the right stance.

Dinosaur modelling is a skill fully understood by only a few dozen people worldwide. The modeller must start with the palaeontologist's bone and muscle reconstructions to get the basic body

NATURE VOL. 324 27 NOVEMBER 1986

Baryonyx, a remarkable new theropod dinosaur

Alan J. Charig & Angela C. Milner

Department of Palaeontology, British Museum (Natural History), Cromwell Road, London SW7 5BD, UK

An extremely large claw bone, some 30 cm long, was found in Wealden (Lower Cretaceous) deposits in a Surrey claypit in January 1983. This led to the discovery the following month of the well-preserved skeleton of a new large theropod dinosaur. Only one other theropod specimen comprising more than a few bones had ever been found in Britain, and that discovery was more than a century ago. Indeed, no large theropod, reasonably complete, had previously been discovered in Lower Cretaceous rocks anywhere in the world. Our study so far suggests that the Surrey dinosaur was a typical large theropod in certain respects, resembling, for example _Allosaurus_[1]. In several other respects, however, it differs sufficiently from all known dinosaurs to merit designation as the representative of a new species, genus and family.

Family **Baryonychidae** nov.

Diagnosis. As for genus *Baryonyx*, below.

Genus *Baryonyx* nov.

Derivation of name. Gk, *βᾰρύ*, heavy, strong; *"ονυξ*, talon, claw.

Type-species. *B. walkeri* nov.

Diagnosis. As for species *B. walkeri*, below.

Species *Baryonyx walkeri* nov.

Derivation of name. For Mr William J. Walker, who found the first indication of the skeleton, namely the large claw bone (ungual phalanx).

Holotype. B.M.(N.H.) Palaeo. Dept. R. 9951.

THE *BARYONYX* STORY: IDENTIFICATION. After detailed study, The Natural History Museum palaeontologists decided that the dinosaur was one of a new and unknown species. Its existence was officially announced in November 1986, nearly four years after the original discovery. It was named *Baryonyx walkeri*. *Baryonyx* from the Greek meaning 'heavy claw', and *walkeri* in honour of the man who found the claw bone.

LAST MEAL. *Opposite:* Close examination of this saurischian *Compsognathus* revealed tiny bones of the lizard *Bavarisaurus* inside its ribs, presumably its last meal.

CLADOGRAM. *Overleaf:* This shows some of the features that are used to identify different dinosaurs. By answering a series of 'Does it have?' questions, you can work out the group it belongs to.

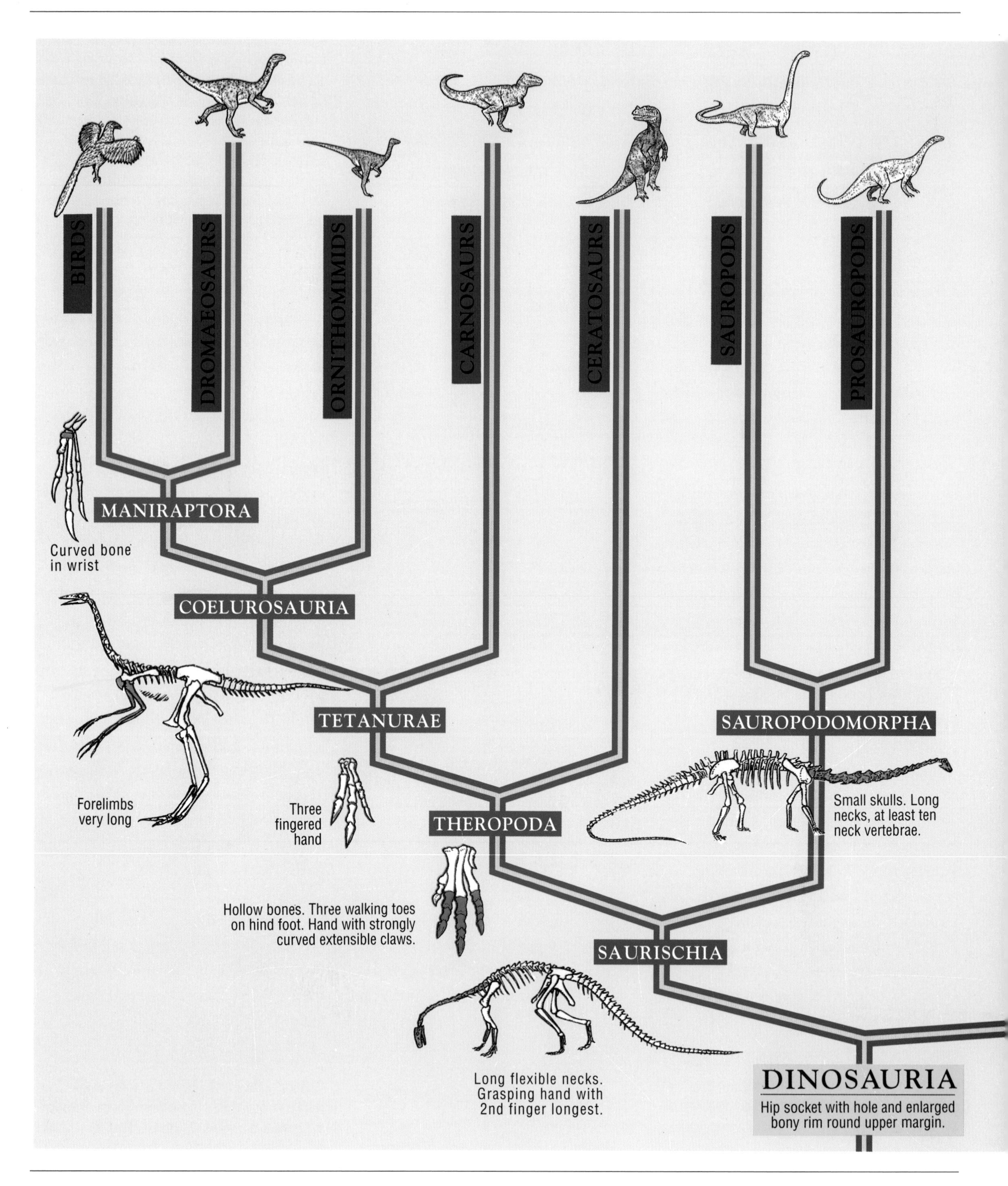
BIRDS
DROMAEOSAURS
ORNITHOMIMIDS
CARNOSAURS
CERATOSAURS
SAUROPODS
PROSAUROPODS
MANIRAPTORA
Curved bone in wrist
COELUROSAURIA
Forelimbs very long
TETANURAE
Three fingered hand
SAUROPODOMORPHA
Small skulls. Long necks, at least ten neck vertebrae.
THEROPODA
Hollow bones. Three walking toes on hind foot. Hand with strongly curved extensible claws.
SAURISCHIA
Long flexible necks. Grasping hand with 2nd finger longest.
DINOSAURIA
Hip socket with hole and enlarged bony rim round upper margin.

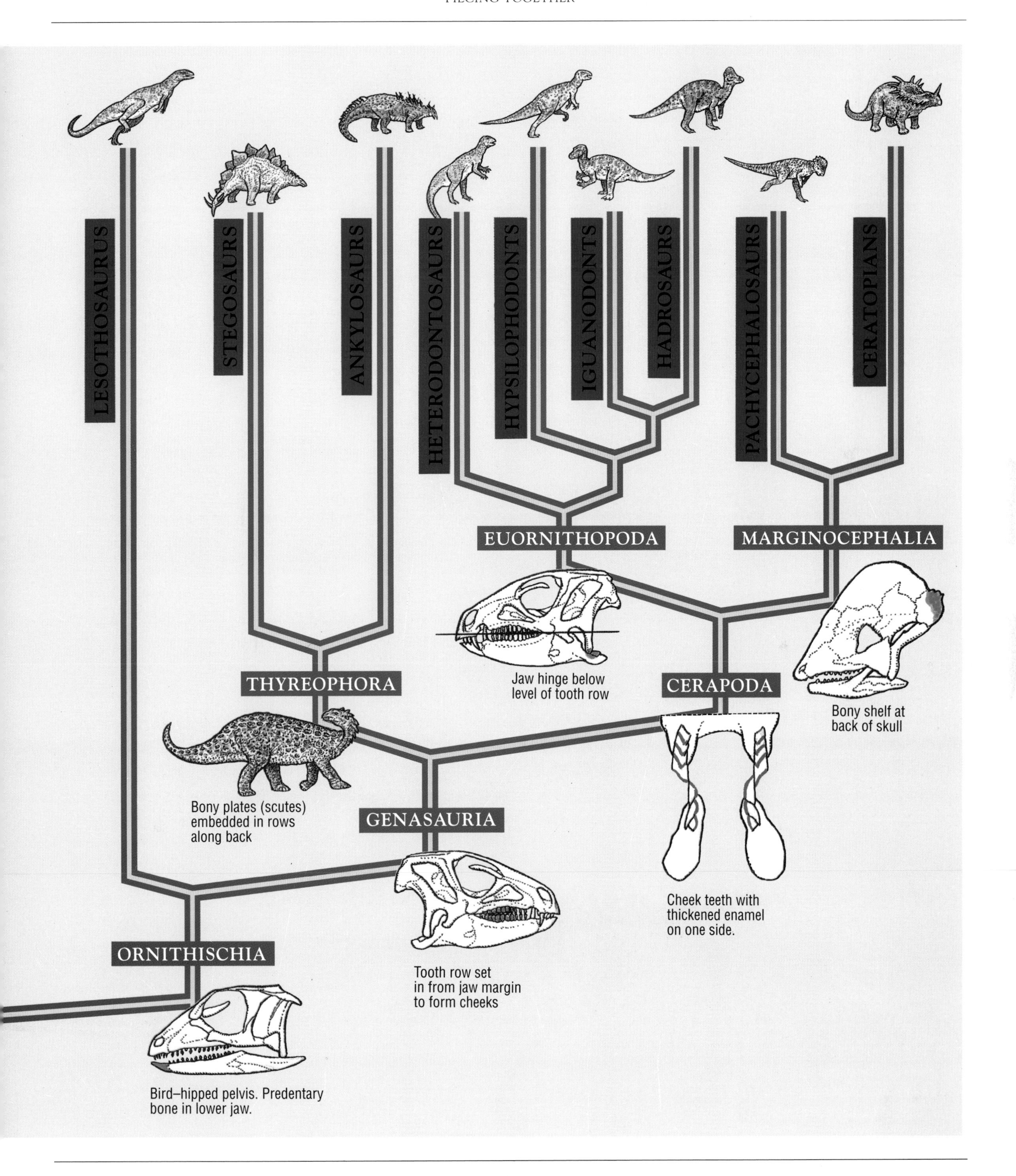
LESOTHOSAURUS
STEGOSAURS
ANKYLOSAURS
HETERODONTOSAURS
HYPSILOPHODONTS
IGUANODONTS
HADROSAURS
PACHYCEPHALOSAURS
CERATOPIANS
EUORNITHOPODA
MARGINOCEPHALIA
Jaw hinge below level of tooth row
THYREOPHORA
CERAPODA
Bony shelf at back of skull
Bony plates (scutes) embedded in rows along back
GENASAURIA
Cheek teeth with thickened enamel on one side.
ORNITHISCHIA
Tooth row set in from jaw margin to form cheeks
Bird–hipped pelvis. Predentary bone in lower jaw.

shape of the dinosaur. Details such as teeth and claws are again based on the original description. Fossil claws, for example, are in fact only the bone core of the claw itself and need to be modelled up to a third larger. Skin texture added to the model is often based closely on the appearance of modern reptiles such as crocodiles as there is rarely any fossil skin to guide the model maker. Other details are not so clear cut. What colour should the skin be? Mouth open or closed? Should the eyes have round pupils like a bird or slit ones like a snake? There are no right answers to these questions, so the palaeontologist and model maker must decide between them. If the dinosaur model or picture is to have a scenic setting, then the correct vegetation and landscape must also be created.

THE BARYONYX STORY: RECENT DEVELOPMENTS

Baryonyx, whose story we have been following through this chapter, was the first of its kind to be found anywhere in the world. Since it was named in 1986, *Baryonyx* has been the key to understanding and interpreting a series of exciting discoveries. This is how palaeontology progresses, with an ever-growing database in scientific papers providing the groundwork for evaluating new finds and comparing them to what is known already.

Baryonyx was based originally on just one individual, represented by around 60% of the skeleton, as we saw earlier in this chapter. One of its many distinctive features are its teeth, which are not like those of typical meat-eating dinosaurs described in Chapter 3. In keeping with its crocodile-like jaws, the teeth of *Baryonyx* are rather straight, rounded rather than blade-like, and have minute serrations along the front and back edges that are barely visible to the naked eye. They are so crocodile-like that they have been mistaken for crocodile teeth in the past. Careful scrutiny of dozens of isolated crocodile teeth in the collections of The Natural History Museum have turned up several from early Cretaceous localities in Surrey and Sussex that match *Baryonyx*. The palaeontologists who identified and catalogued those teeth more than a century ago could not possibly have worked out their true identity without the important clues provided by the skull and jaws. Neither did they have modern tools like scanning electron microscopes to study the teeth in detail. This all goes to show how vital museum collections are to the study of dinosaurs. We might not realise the significance of an odd bone or tooth now, but it might turn out to be important in the future when new fossils and new techniques come along. Isolated teeth of *Baryonyx* and a few bones have since turned up on the Isle of Wight too. So, although its remains are still very rare (much rarer than those of *Tyrannosaurus*), we can safely deduce that *Baryonyx* lived and fished across the flood plains, swamps and marshes that were southeast England around 124 million years ago. In early Cretaceous times, parts of Kent, Surrey, Sussex and Hampshire were occupied by a large lake of fresh to brackish water, the Wealden Lake. Dry land lay to the north where London is today; to the south lay the Anglo-Paris Basin and the open sea. Two large rivers flowed down from the north and northeast to discharge their waters into the lake through a common delta, with shallow creeks and oxbow lakes. The climate was, in modern terms, subtropical.

The fossil flora and fauna of the locality where *Baryonyx* was found give us some idea of the flourishing environment in which it lived. Common plant remains include ferns, an aquatic or marsh-dwelling herbaceous plant that grew in vast stands, horsetails, club mosses and conifers. Insects were abundant and different kinds of small crustaceans and bivalves lived in the water. Vertebrates included sharks, bony fishes (notably *Lepidotes* on which *Baryonyx* fed),

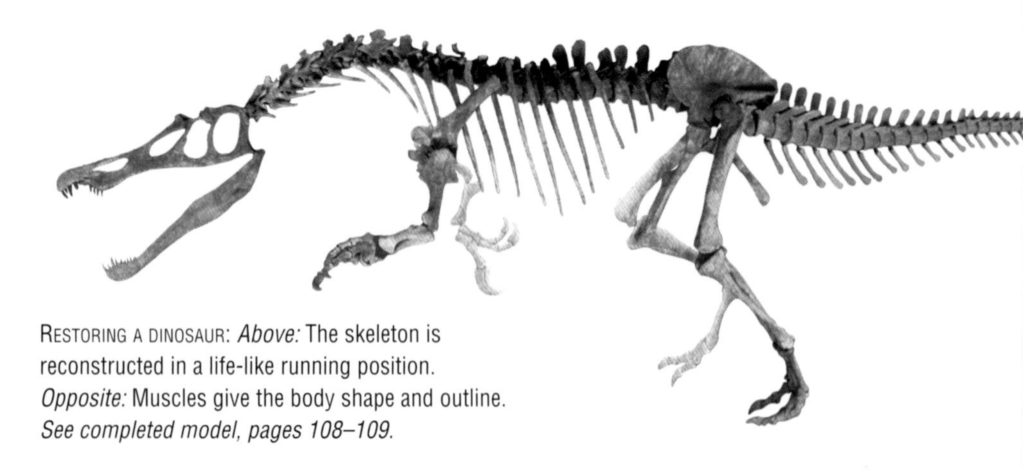

RESTORING A DINOSAUR: *Above:* The skeleton is reconstructed in a life-like running position. *Opposite:* Muscles give the body shape and outline. *See completed model, pages 108–109.*

crocodiles and pterosaurs. Apart from *Baryonyx*, the only other dinosaur remains found in this locality are a considerable number of bones of the plant-eater *Iguanodon*, and a very few isolated bones of small sauropods. Nearby localities of the same age provide a wider picture of dinosaur life in early Cretaceous southeast England: an armoured dinosaur, *Polacanthus*; several sauropods including a brachiosaur; large and small theropods and ornithopods, particularly *Iguanodon*, have all been found after 150 years of collecting. *Baryonyx*' remains were preserved in fine silty clays laid down in an area of shallow water which neither dried out nor flooded. The rocks surrounding the bones of *Baryonyx* suggest that its carcass was unlikely to have been carried very far by the water. It probably died close to were it was buried — it might even have been mired in a soft mud as it crossed the river delta. It certainly did not die of old age; it is clear from the lack of fusion between some of the bones that it was not a fully mature adult. The excellent preservation of most bones, which show no signs of predation or scavenging, suggests that the carcass must have been covered with sediment fairly rapidly. Total decomposition of soft tissues resulted in the disarticulation of the bones. Weathering patterns on some of them suggest intermittent surface exposure of parts of the skeleton, either by receding water levels or by shifting of fine sediment. Some bones had been broken or split before they fossilized. The left lower jaw had been snapped in two but was still partly connected, with the pieces sticking up in a 'V' shape. It was just as if it had been trodden on. Indeed, these odd breaks suggest that the skeleton had been trampled over by large animals while the bones were buried in shallow sediment. A herd of *Iguanodon* or a family of sauropods might have wandered over the spot. The fossil record sometimes provides clues as to how or why a dinosaur died. As we have seen, careful study of the nature of the entombing rocks and the way the bones are preserved inside them, called taphonomy, can reveal much about a dinosaur's postmortem fate.

Baryonyx lived not only in England but in northern Spain too. An upper jaw fragment from La Rioja province turned up in 1995. In addition, two snout ends from slightly younger rocks in Niger in Africa have proved to be indistinguishable from *Baryonyx*. The French palaeontologist, Philippe Taquet, who described them in 1984, was so puzzled as to their identity (because there was nothing else remotely like them at the time) that he did not give them a scientific name. Now we know that *Baryonyx* and very close relatives lived in northern Africa as well as Europe.

As detailed work on *Baryonyx* was completed it became clear that it had some features in common with an even more bizarre dinosaur called *Spinosaurus*. *Spinosaurus* was found in Egypt in 1901 and described by German palaeontologist Ernst Stromer in 1915. Although known only from parts of one skeleton, it was a huge animal, at least as big as *Tyrannosaurus*. The next task then, would have been to compare the bones of *Spinosaurus* with *Baryonyx*. Sadly, that can never be done. *Spinosaurus* was housed in Munich Museum and was destroyed in 1944, during World War II, leaving only Stromer's description and figures. Tantalizingly those drawings show that the front part of the lower jaw is similar in shape to *Baryonyx*. The teeth are

similar too, conical, uncompressed and sharp-edged, but unserrated in this case. Some of the vertebrae were also very like those of *Baryonyx*. However, the dorsal vertebrae of *Spinosaurus* have enormously elongated back spines, almost a metre long, so it must have had either a 'sail' running along the back, the skin stretched between the spines or a large fleshy hump. In contrast, *Baryonyx* had short-spined vertebrae and no sail. How closely related are *Baryonyx* and *Spinosaurus*? New material has come to the rescue to help answer that question. A dinosaur, named *Suchomimus*, was discovered by American palaeontologist Paul Sereno in 1998 from the remote Ténéré Desert in Niger, the same place as Philippe Taquet's mysterious snout ends. *Suchomimus* is very like *Baryonyx* indeed, although it is a few million years younger, about 30% larger, with slightly elongated back spines. The two were certainly closely related. Thousands of isolated *Spinosaurus*-like teeth have turned up at Cretaceous dinosaur sites all over northern Africa, from Morocco in the west to Egypt in the east, and as far south as Cameroon and northern Kenya. These sites are geologically a bit younger again — some 98 to 95 million years old. Frustratingly, remains of skulls and skeletons are much more rare and very fragmentary, but some bones have turned up — vital clues in the detective story.

A huge lower jaw of *Spinosaurus* from Morocco in The Natural History Museum collections closely matches Stromer's original drawings and shows many features in common with *Baryonyx*. A large fragmentary snout from Morocco also compares closely with *Baryonyx*. Did both the jaw and the snout belong to the same kind of animal? The matching morphology of bones and teeth confirm that they did, although they were from individuals of different sizes. The shape of the snout is very similar to *Baryonyx*, although it is even more stretched towards the tip. The animal certainly had very long slender jaws, with an S-shaped curve and expanded tip just like *Baryonyx*. Although research is not yet finished, it is clear that *Baryonyx*, *Suchomimus* and *Spinosaurus* are all closely related animals that share jaw and tooth characters that define a group of long-snouted theropod dinosaurs which take their name — spinosaurs — from the original *Spinosaurus*. The jaw morphology of *Baryonyx* suggested to Alan Charig and Angela Milner the possibility of fish-eating and/or scavenging habits. Both these ideas are corroborated by gut contents — fish — as we have seen already, and also some bones of a very small *Iguanodon*. *Spinosaurus* exhibits an even more specialized condition in an animal that was perhaps twice the size of *Baryonyx*. *Spinosaurus* must have had a skull nearing 2 metres long and the whole animal was at least 15 metres long.

Spinosaurs are highly-specialized long, slender-snouted theropods whose jaws and teeth suggest a fish-eating diet. Their remains are found in low-lying flood plain, lake or coastal areas together with abundant fish fossils. This is likely to have been their natural habitat. In North Africa, where *Spinosaurus* must

CROCODILE MIMIC. The teeth of *Baryonyx* are conical and rounded with a very finely serrated cutting edge seen in the scanning electron microscope picture (*left*). Crocodile teeth are very similar except that they have a smooth cutting edge.

CLOSE RELATIVE. *Suchomimus tenerensis*, 'the crocodile mimic from the Ténéré Desert' was a close African relative of *Baryonyx*. It was about 30% larger and lived about 5 million years later than *Baryonyx* in Niger, central West Africa.

have been very common judging from the huge numbers of teeth found, fossils of giant lungfishes and coelacanths, 4 to 5 metres long, are found in the same beds. Those giant sail-backed fishermen would have had a plentiful food supply which no other dinosaurs were equipped to tackle. There is still much to learn about spinosaurs, but ten years ago we hardly knew they existed, nor what a major role they played in the ecology of coastal flood plain ecosystems.

Where did spinosaurs originate? To complicate the biogeographical picture, two more new spinosaurs, *Irritator*, a skull lacking the snout, and *Angaturama*, a snout lacking a skull, turned up in Brazil in 1966 from rocks about 110 million years old. Since the earliest spinosaur, *Baryonyx*, occurs in Europe, the simplest interpretation of their geographical distribution is that the group originated in the northern supercontinent of Laurasia. They dispersed via an intermittent Iberian land bridge into the southern continent of Gondwana. There were no barriers to moving south, east as far as Egypt and west into Brazil, before South America and Africa began to move apart as the opening of the Atlantic Ocean began. Thereafter, spinosaur populations were isolated from each other and different forms evolved in Africa and South America.

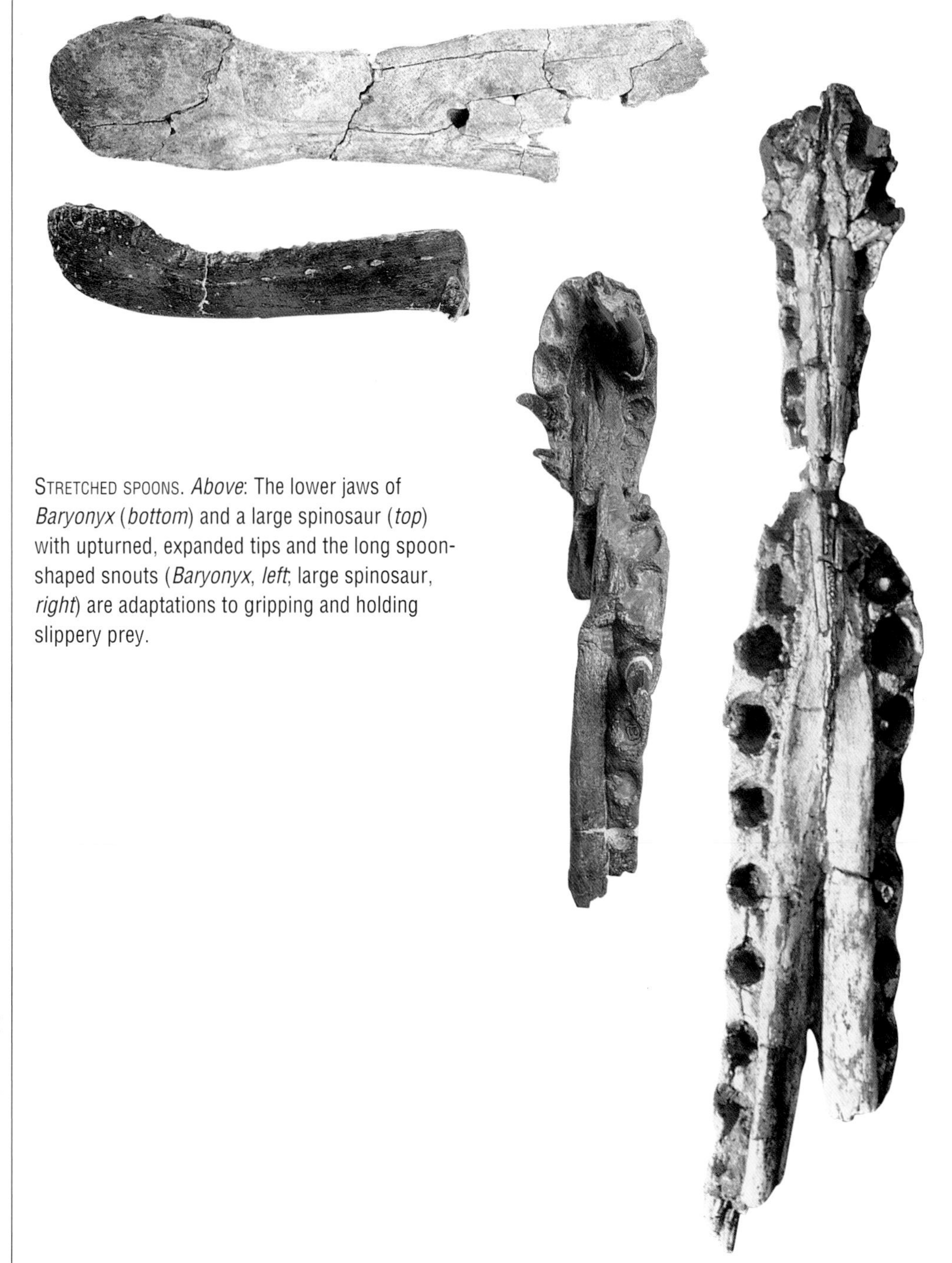

STRETCHED SPOONS. *Above*: The lower jaws of *Baryonyx* (*bottom*) and a large spinosaur (*top*) with upturned, expanded tips and the long spoon-shaped snouts (*Baryonyx, left*; large spinosaur, *right*) are adaptations to gripping and holding slippery prey.

DINOSAURS

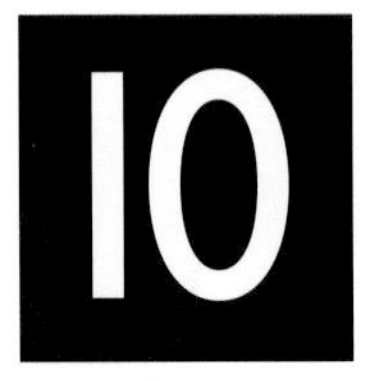

The story of dinosaurs does not end with the extinction of *Tyrannosaurus rex* and *Triceratops* 65 million years ago. They did not all die out — one group of small meat-eaters

AND BIRDS

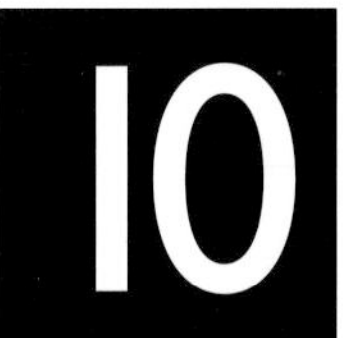

gave rise to all the birds that we see around us now. Spectacular new fossils from China have provided conclusive evidence that birds are small feathered dinosaurs.

Are birds really the descendants of meat-eating dinosaurs? How could it possibly be that a small delicate bird such as a robin could have descended from such a lumbering giant like *Tyrannosaurus?* Birds and dinosaurs seem so different. Sufficiently different that birds have traditionally been classified separately as feathered and (mostly) flying animals. Spectacular new feathered fossils discovered in China have proved to be the final clues in a long detective story that began 150 years ago. They also fulfil predictions made by many modern palaeontologists. If birds are the descendants of meat-eating dinosaurs then those dinosaurs must have had feathers too. They did — and the Chinese fossils prove it beyond all doubt. They also tell a fascinating story of why feathers first evolved — and it was certainly not for flight.

It has been widely accepted for more than 30 years that birds are direct descendants of small maniraptoran theropods. This followed American palaeontologist John Ostrom's discovery of *Deinonychus*, a 3 metre-long, lightly built, fast-running predatory dinosaur. Found in 110 million-year-old early Cretaceous rocks in Montana, USA, it belongs to a family of advanced theropods called dromaeosaurs, the 'raptors', which also includes *Velociraptor* from the Gobi Desert in Mongolia. *Deinonychus* showed a novel feature in the hand skeleton, a half-moon shaped wrist bone called a semi-lunate carpal. This allowed the wrist to be flexed sideways in addition to up and down movements. Dromaeosaurs could fold their long hands almost in the manner that birds do today. Another advantage was that the hands could be swivelled and rapidly whipped forwards to grab prey. It is no coincidence that the movement is similar to the flight stroke of a bird. This ability to rotate the wrists was shared with other small advanced theropods — oviraptorids and troodonts. All three families of long-handed theropods are collectively called maniraptorans (see cladogram on pages 118–19).

The semi-lunate carpal is shared uniquely with the earliest known bird, *Archaeopteryx*, discovered more than 150 years ago from 147-million-year-old latest Jurassic rocks in Germany (see page 18). Many other features of *Archaeopteryx*' skeleton compare closely with maniraptorans. More impressive still are the impressions of wing and tail feathers, imprinted in the fine-grained rock surrounding the *Archaeopteryx* skeleton. The wings bear asymmetric primary and secondary flight feathers that are almost identical with those of

A CLUE IN THE WRISTS. *Deinonychus* and *Archaeopteryx* shared many similarities, particularly the half-moon-shaped wrist bones that permitted sweeping and folding movements of the hands.

(Previous spread) *SINOSAUROPTERYX* RESTORED. This small, fast-running predator had a simple, hollow, feather-like coat, which provided insulation.

modern birds. The midrib or quill of asymmetric feathers is offset from the middle of the feather vane so that there is a short leading edge and a longer trailing edge. This controls the flow of air over the wing surfaces, aerodynamic properties necessary for flight. So, *Archaeopteryx* represents a snapshot of evolution 'caught in the act', a dinosaur equipped with wings and capable of flight — an ancient bird about the size of a modern magpie. Somewhere along that evolutionary line feathers — the feature that defines birds — must have first appeared in theropod dinosaurs but the fossil record did not preserve such information. Now that gap has been filled with the recent discovery of feathered fossils that offer a whole new perspective on the origin of birds.

CHINESE FOSSIL TREASURES

A huge array of exquisitely and exceptionally preserved fossils have been discovered from a series of early Cretaceous volcanic deposits in the Sihetun region of Liaoning Province in northeastern China over the last 15 years. Beds of rock dated at between 130 and 125 million years old, and consisting of a series of volcanic basalts alternating with fine-grained volcanic ash and tuffs were laid down in a forested lakeside setting in a volcanically active area. Literally millions of plants, insects, molluscs, crustaceans, fishes, frogs, salamanders, turtles, lizards and lizard-like swimming animals, mammals, birds and dinosaurs were suffocated, entombed rapidly in volcanic dust and ashes. Many ended up on the lake bed flattened and compressed in the sediments. The dust and ash sealed out oxygen, preventing decay and allowing the preservation of keratinous soft structures such as skin, hair and feathers. Such detailed preservation is very rare indeed. Others were buried alive and preserved as 3D fossils but soft tissues were not preserved — literally a dinosaurian Pompeii. One 3D skeleton discovery, *Mei long*, a small troodont (a maniraptoran), was found in a sleeping position, crouching on the ground, its tail curled around its body and its head tucked under its forelimb — a position to conserve body heat, just like a sleeping bird. Both these preservations provide a vivid picture of the rich variety of life in that corner of Asia in early Cretaceous times.

A large-scale industry set up by local farmers has led to the excavation of vast numbers of fossils. They include thousands of specimens of the commonest Liaoning bird, *Confuciusornis*, which must have lived in large flocks in the trees around the margins of contemporary lakes and probably succumbed to poisonous volcanic gases and ash falls. Much more rare are a series of spectacular small, feathered dinosaurs — 'dino-birds' that represent a snapshot of a wide span of dinosaur evolution.

FEATHERED THEROPODS

The first Liaoning dino-bird, *Sinosauropteryx*, discovered in 1995, created a huge sensation. Here at last was what palaeontologists had been waiting for; the predictions about feathered dinosaurs had come true and the last piece could be fitted into the jigsaw puzzle of bird origins. *Sinosauropteryx* was very clearly a small two-legged predator. It had toothed jaws equipped with flattened serrated teeth, the pattern typical for meat-eating dinosaurs. Its forelimbs were quite short and ended in clawed

EARLY FEATHERS. The dark structures running from head to tail on this juvenile *Sinosauropteryx* are the remains of a covering of simple hollow 'downy' filaments or 'protofeathers'.

fingers. The hind limbs, especially long from the knee downwards, were typical of fast runners. A juvenile *Sinosauropteryx* measuring 55 centimetres long from nose to tail and adult skeletons tell us that it grew to at least 150 centimetres long. One adult also shows us what it had eaten for its last meal — the lower jaw bones of a mammal are still in position near the end of the gut. *Sinosauropteryx* probably snapped up anything small enough to catch such as lizards and mammals. Its outer covering, preserved best on the juvenile specimen, does not look like the branched feathers that we are familiar with. Rather it seems to consist of single hollow feather-like fibres that may represent an early simple stage in the origin and development of feathers. These discoveries enable predictions to be made about how far back down the family tree of dinosaurs feather-like coverings appeared. If coelurosaurs had feathers, then so did all the groups that evolved after them — and that includes tyrannosaurs, oviraptorids and dromaeosaurs.

So why did feathers arise? Insulation is the most likely explanation. A coat of hollow filaments would be useful in helping to keep heat in — especially in small-bodied and young animals. The simple physical ratio of surface to volume means that small bodies have more surface area relative to their volume and so lose (or gain) heat more rapidly. It might seem very far-fetched, however, to envisage a feathery *Tyrannosaurus*. A six tonne beast had a great deal of bulk to buffer temperature changes and is unlikely to have sported feathers, but its chicks (as yet unknown) could have been as fluffy as the farmyard version to keep them warm while they grew quickly. There are a few skin impressions of *Tyrannosaurus* and they show that it was scaly; feathers may have been shed as it grew large. We do know, thanks to Liaoning, that tyrannosaurs certainly did have feathers. *Dilong*, a small and primitive tyrannosaur about 1.6 metres long discovered in 2004 had downy, filamentous 'protofeathers', similar to those of *Sinosauropteryx*, preserved near the back of the jaws and around part of the tail.

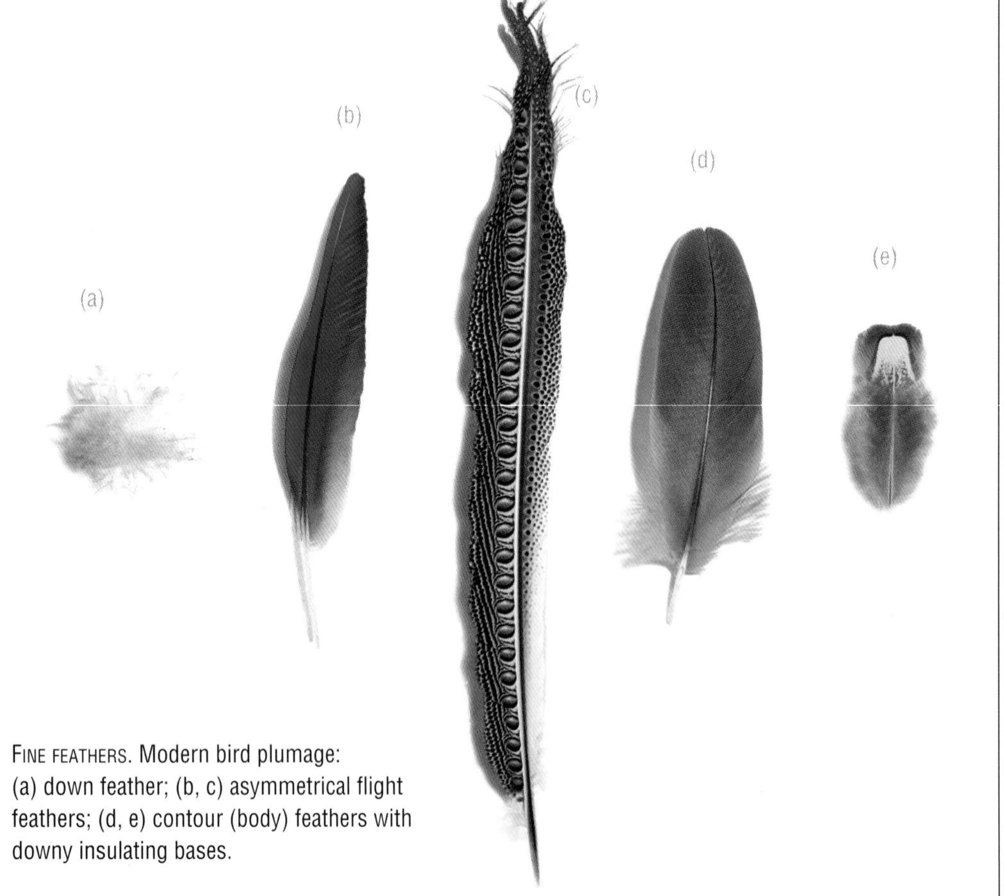

FINE FEATHERS. Modern bird plumage: (a) down feather; (b, c) asymmetrical flight feathers; (d, e) contour (body) feathers with downy insulating bases.

Modern bird feathers consist of quill embedded in the skin that emerges as a shaft from which the feather sprouts. Flight feathers have a strong rigid tapering shaft supporting the feather vane. The vane is formed from a series of barbs interlocked together into a continuous sheet by hook-like barbules. The barbs are asymmetrically arranged along either side of the shaft to give an aerofoil shape with leading and trailing surfaces to produce lift. Down and contour feathers that cover the body are symmetrical with a slender shaft and do not have the aerodynamic qualities needed for flight. Richard Prum, at the University of Kansas, has proposed a model of how feathers may have originated. Feather follicles first gave rise to simple hollow sheaths that gradually became more complex over time to produce increasingly elaborate feathers, down-like ones first and then, by different pathways, feathers with parallel barbs and barbules, and eventually asymmetric vanes. Recent embryological studies of how feathers form match the developmental model and so do the fossils.

Feathers may have had an important role in the development of brooding behaviour. Bird-like brooding behaviour is known from a dramatically preserved small maniraptoran, *Oviraptor*, from the Gobi Desert. As seen in Chapter 5, it was preserved squatting on its nest of eggs — dinosaur brooding behaviour frozen in time for 75 million years. Here was dramatic evidence that birds have inherited the nesting habits of their dinosaur ancestors, habits that probably go even further back in time than the *Oviraptor* discovery. Equally clearly, large dinosaurs could not have sat on their

FROZEN IN TIME. *Caudipteryx* was preserved in a typical death pose with the neck bent backwards as the ligaments contracted. This adult individual shows the short arms fringed with feathers, a gizzard filled with gastroliths, long, strong hind limbs and peacock-like tail plumes.

eggs. There is, as yet, no evidence of what *Tyrannosaurus* did, but fossil nests of some large plant-eating duck-billed dinosaurs or hadrosaurs demonstrate an alternative way of incubating eggs — covering them with sediment and vegetation to control the temperature, as we saw in Chapter 5. Crocodiles also do that by building nest mounds and some birds called megapodes do too. That demonstrates just one of the links between birds and dinosaurs and crocodiles, they share many anatomical features that are not found in any other living animals. On that basis, the two are more closely related to each other than either is to any other living animal group. The fossil record of dinosaurs shows all the evolutionary reproductive changes that took place between crocodiles and birds.

FEATHERS, FANS, PLUMES AND FLUFFY COATS

Feathered dinosaur discoveries from Liaoning have come thick and fast since those initial discoveries of *Sinosauropteryx*. They show in fantastic detail how feathers became elaborated and adapted for different functions. *Protarchaeopteryx* is known from a single, frustratingly incomplete individual first described in 1997. It was clearly a long-legged running theropod with no suggestion of a backward pointing toe. It has very long arms and huge sharply clawed hands, almost as long as those of *Archaeopteryx* and those swivelling wrists for seizing prey. *Protarchaeopteryx* had a short tail with a clump of feather at the tip. These are true feathers with a central shaft and symmetrically arranged barbs. A few scattered feathers near one of the forelimbs show the same structure. But they are not flight feathers like the asymmetrical ones possessed by *Archaeopteryx*. *Protarchaeopteryx* was not a flier; those long arms long arms with swivelling wrists had evolved first for capturing prey. *Protarchaeopteryx's* tail feathers might have been used for some other function — display perhaps?

Caudipteryx, also named in 1977, provides a dramatic example of feathers for show, a jaunty fan of feathers on the end of its tail. Numerous specimens, all about the size of a turkey — probably adult size, have been found from the Sihetun area. *Caudipteryx* was a very long-legged runner with powerful hind

A UNIQUE FIND. *Protarchaeopteryx* was a long-legged turkey-sized predator. Its arm and tail feathers might have been used for display. It is known only from one specimen.

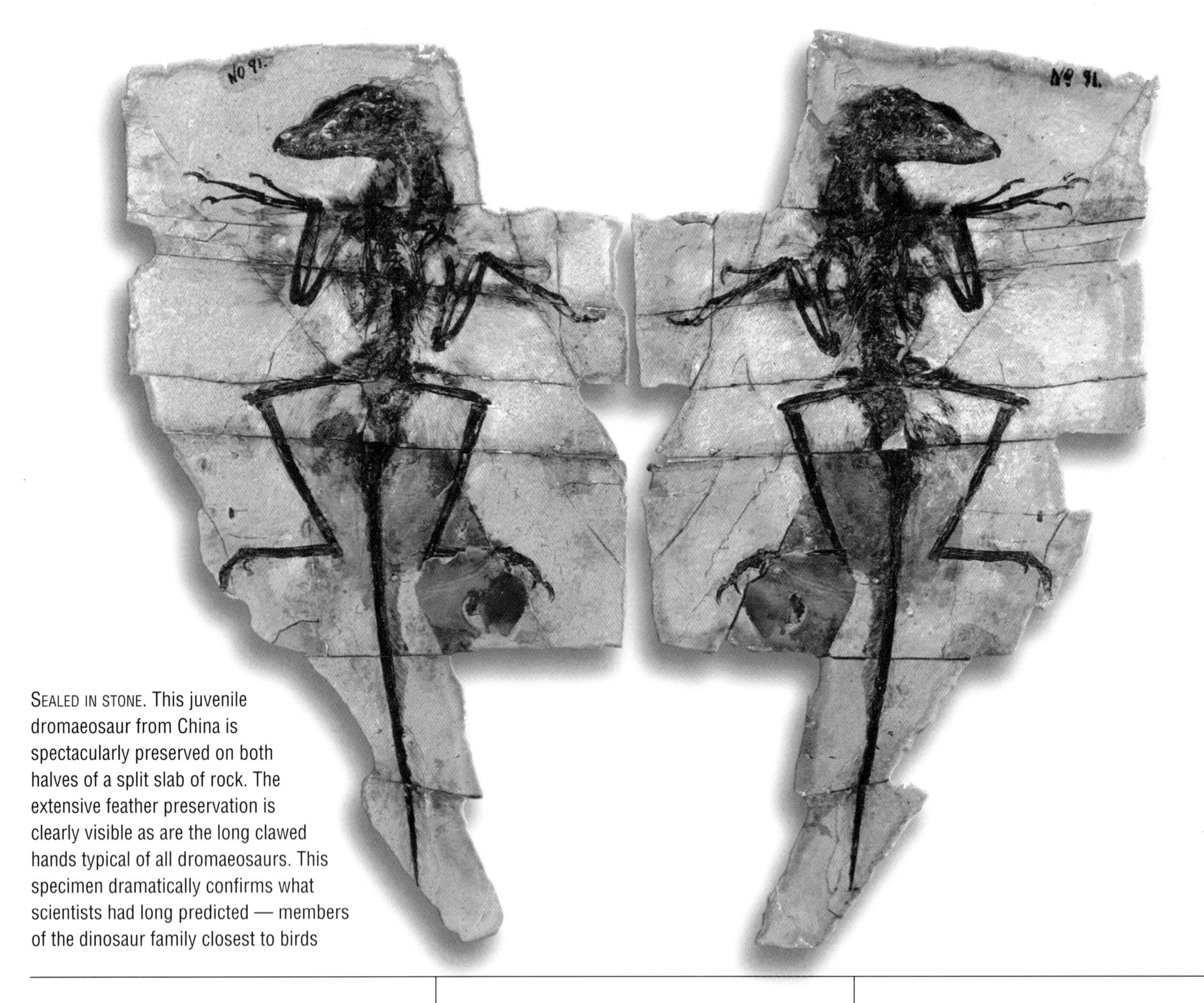

SEALED IN STONE. This juvenile dromaeosaur from China is spectacularly preserved on both halves of a split slab of rock. The extensive feather preservation is clearly visible as are the long clawed hands typical of all dromaeosaurs. This specimen dramatically confirms what scientists had long predicted — members of the dinosaur family closest to birds

limbs and unusually short arms tipped with claws. It also had quite a short skull with just a few small teeth at the tip of the upper jaws and a beak. A heap of stomach stones (gastroliths) is preserved in most individuals, a gastric mill in life that functioned to grind up hard food, aided by muscular walls of the gizzard — the powerful grinding 'stomach'. Farmyard chickens pick up grit to do the same job. So what kind of lifestyle can we deduce for *Caudipteryx*? A fast runner with a diet that included hard food gathered with a beak — it could have eaten fruits and seeds from the lush vegetation around the lakeside, or fed on shelled crustaceans and molluscs that lived in the lake shallows.

Caudipteryx has beautifully preserved long symmetrical feathers attached to its very short forelimbs. Those arms were impossibly short for any thoughts of flying so what were the feathers for? Perhaps they could have been used for display in conjunction with the tail. *Caudipteryx* is thought by some palaeontologists to be related to the beaked oviraptorids — like that Mongolian nesting *Oviraptor* with its arms curled around its eggs. Its arms may well have been feathered too, but they were not preserved in the coarse desert sands.

Feathered dromaeosaurs have now been found from Liaoning. *Sinornithosaurus* dating from 1998, confirmed that the family closest to *Archaeopteryx* were feather-covered, which is just what we would expect to find. Perhaps the most stunning dino-bird of them all is a juvenile dromaeosaur discovered in 2000. It is covered from head to tail in an insulating coat of fine, branched feathers,

FUZZY RAPTOR. This juvenile dromaeosaur (see specimen opposite) was covered in a coat of branched feathers but did not possess any flight feathers.

a longer fringe of feathers along the arms, and is even clad in long rook-like 'trousers'. This individual, affectionately known as 'Dave the fuzzy raptor', may be a juvenile *Sinornithosaurus*. However, it was not fully-grown so it is not possible to be sure whether some differences in the skeletons are simply due to size. We can be sure that it did not fly, for there is so sign of asymmetric flight feathers.

THE ORIGIN OF FLIGHT

Liaoning's amazingly preserved dino-birds lead to the inescapable conclusion that dinosaurs did indeed possess feathers and that they evolved initially for purposes other than flight. It seems likely that a simple insulating cover arose first and was later modified for display, signalling and finally flight. So, how did dinosaurs get off the ground and learn to fly?

Did flight begin from the trees down or from the ground up? This has been hotly debated since the discovery of agile, fast-running dromaeosaurs such as *Deinonychus*. The 'trees down' hypothesis holds that flight evolved from an arboreal gliding stage. This would require the ability to climb tree trunks and the ability to glide from tree to tree or tree to ground. The 'ground up' hypothesis suggests that flight evolved from small ground-dwelling theropods with running adaptations. It requires that wing feathers were co-opted for flight by functioning first as insect trap nets. The animals would chase small prey using sweeping motions of the forelimbs — as we have seen in the dromaeosaur forelimb — permitted by modifications in the shoulder girdle, wrist and hand. This motion together with running speed would eventually generate sufficient lift and thrust for the runner to become airborne — almost by accident. This idea became widely accepted during the 1980s and 1990s but the discovery of a tiny Liaoning dromaeosaur, *Microraptor*, with arboreally adapted foot claws has swung the debate back into the trees.

The smallest dromaeosaur from Liaoning, *Microraptor*, provides a test of the arboreal habits of these animals. With a body length of just 47 millimetres, *Microraptor zhaoianus* is the smallest adult non-avian dinosaur known. It was also feathered, traces of long branched feathers were preserved attached to the forelimbs and legs. Its toes, as well as its fingers, were

CLIMBING CLAWS. *Microraptor*'s very curved hand and foot claws strongly suggest that it would have been a capable tree-climber.

equipped with slender and very curved claws; the degree of claw curvature falls strictly within the range of modern climbers. The Chinese scientists who described *Microraptor* in 2000 suggest that it was a tree-dweller not a ground runner — arboreal rather than cursorial.

Even more surprising was the discovery of a second species of *Microraptor*, M. *gui*, found in 2003. It has a few asymmetrical feathers and a very wing-like feather configuration in the forelimb, which might represent a late stage in the transition from a gliding structure to one capable of true flight. Amazingly, the single known specimen of M. *gui* also sports long asymmetrical flight feathers on its hind limbs. This has given rise to speculation as to whether there might have been a four-winged gliding stage in flight evolution with the hind wings acting like stabilising canards on an aircraft. Interestingly, a four-winged stage was suggested as far back as the 1930s as a purely theoretical hypothesis. Whether flight really began this way or whether *Microraptor* was an experimental sideline must await further fossil discoveries.

Recent studies on escape behaviour in modern ground-living birds such a quail and partridge provides an interesting addition to the debate. They employ 'wing-assisted vertical running' to get off the ground, beating their wings rapidly to generate a down force to help them stick to the substrate while climbing a bush or tree trunk. Even the fluff on chicks' wings increases the wing surface area sufficiently to allow efficient climbing. Thus it has been suggested that dino-bird arm feathers could have functioned in the same way — as downforce generators and 'proto-wing' gliding structures. The prey-catching idea is now giving way to an hypothesis that the selective pressures leading to flight may have arisen as a predator escape mechanism in small theropods whereby the presence of feathered arms would aid rapid climbing away from danger and allow gliding from perch to perch or perch to ground. *Archaeopteryx* represents a late stage in this process with a modern wing configuration of asymmetric primary and secondary flight feathers permitting limited stable powered flight.

ANCIENT BIRD. *Archaeopteryx* wing structure and a number of asymmetric flight feathers were almost identical to those of modern birds. However, it still retained the dromaeosaurid hallmark of three long, clawed fingers on each hand.

Analyses of the claws of *Archaeopteryx* have interpreted them as multipurpose climbing and walking structures although the claw sheaths show no signs of wear as might be expected in a ground-dweller. The brain and inner ear of *Archaeopteryx* have been studied recently by one of the authors (Angela Milner) using computed tomography, 3D x-rays that can see right inside a fossil skull. This has shown that the brain lobes were very like those of modern birds with enlarged areas dealing with sight and flight co-

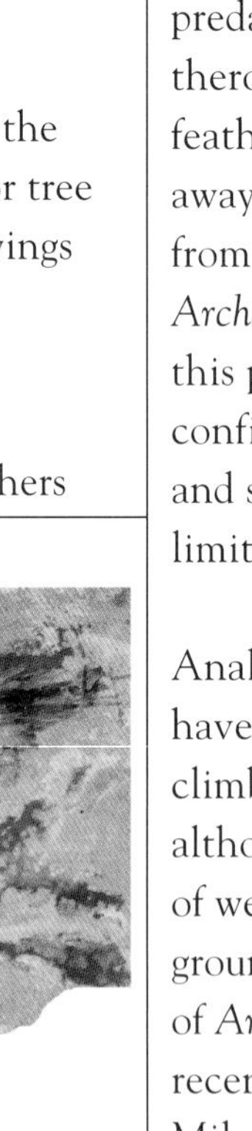

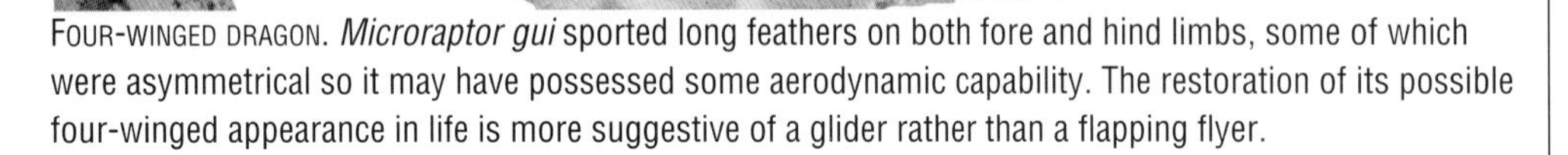

FOUR-WINGED DRAGON. *Microraptor gui* sported long feathers on both fore and hind limbs, some of which were asymmetrical so it may have possessed some aerodynamic capability. The restoration of its possible four-winged appearance in life is more suggestive of a glider rather than a flapping flyer.

ordination. The inner ear is just like that of a modern bird too, with very large semi-circular canals, the organs of balance. These features suggest that *Archaeopteryx* had all the control systems in place that are necessary for flight in keeping with its wing and flight feather structure.

FLIGHT IMPROVEMENTS

True birds were also present in the Liaoning fauna and highlight the diversification of birds and rapid evolution of post-*Archaeopteryx* stage flight apparatus in the early Cretaceous. The majority belonged to a dominant Mesozoic-only group called enantiornithines. *Confuciusornis* had a shortened tail ending in a pygostyle (triangular plate), important for reducing weight and providing a controllable attachment for the tail feathers. It also had bony sternal plates for wing muscle attachment, exceptionally large wings, an adaptation to increase surface area and lift, and is the oldest bird known to have possessed a horny beak — loss of teeth was another weight-saving device. Birds belonging to the Ornithurae, the groups that include all modern birds, were also present in the fauna. A seed-eater and fish-eating shore-wader with a keeled breast bone and other skeletal modifications that suggest strong flying capabilities comparable to modern birds, had already appeared by early Cretaceous times.

TIME AND PLACE

The Liaoning fauna has provided us with a snapshot of an ancient community in which different groups of feathered dinosaurs and flying birds were all living alongside each other. How come they were all living together if birds descended from dinosaurs? A further complication is that *Archaeopteryx* is some 27 million years older than the feathered theropods from Liaoning. The key point to remember is that Liaoning reveals what was living at that time in that part of the world, but not when the various groups evolved. The Liaoning animals are representatives of several lines of evolution, some of which had existed for longer than others.

The Chinese dino-birds have given us many clues that suggest the ways in which dinosaurs could have taken flight, but many stages along the way to airworthiness are still hidden in the rocks of Earth's past history. There is still no evidence of exactly how asymmetric flight feathers developed from downy feathers, how long ago it happened and when the number and arrangement of flight feathers in the wing was established. We need some Liaoning-type preservations much earlier in time — before the appearance of *Archaeopteryx* in the latest Jurassic to provide the answers.

FINELY PRESERVED. This *Confusiusornis* shows the horny beak, long folded wings and a shortened tail ending in an elongated pygostyle. Two long tail feathers continue beyond the bottom right edge of the slab of rock. These were tail pennants identifying the specimen as a male.

DINOSAUR DATA

This alphabetical listing provides comparative information on the dinosaurs and other animals mentioned in this book. As well as providing a guide to the pronounciation and meaning of the name, there are details of the group to which each belongs, age, geological period and size. Ages are given as accurately as possible. Where a range is given (e.g. 155–145 million years), it means that the animal lived within that time span, not for all of it. The illustrations shown are not to scale.

*incertae sedis (group uncertain)

Apatosaurus

Albertosaurus
Al-burt-oh-sore-us • ALBERTA LIZARD

GROUP	Saurischia, Carnosauria
AGE	76–74 million years
PERIOD	Upper Cretaceous
LOCALITY	Canada
LENGTH	up to 9 metres

Allosaurus
Al-oh-sore-us • OTHER LIZARD

GROUP	Saurischia, Carnosauria
AGE	150–135 million years
PERIOD	Upper Jurassic
LOCALITY	USA
LENGTH	up to 15 metres

Anchisaurus
Ank-ee-sore-us • NEAR LIZARD

GROUP	Saurischia, Prosauropoda
AGE	200–190 million years
PERIOD	Upper Triassic
LOCALITY	USA
LENGTH	up to 2 metres

Angaturama
An-gat-you-rah-ma • NOBLE

GROUP	Saurischia, Theropoda
AGE	112–98 million years
PERIOD	Early Cretaceous
LOCALITY	Argentina
LENGTH	unknown

Apatosaurus
Ah-pat-oh-sore-us • DECEPTIVE LIZARD

GROUP	Saurischia, Sauropoda
AGE	154–145 million years
PERIOD	Upper Jurassic
LOCALITY	USA
LENGTH	up to 21 metres

Archaeopteryx
Ark-ee-opt-erics • ANCIENT WING

GROUP	Saurischia, Aviales
AGE	147 million years
PERIOD	Upper Jurassic
LOCALITY	Germany
LENGTH	up to 45 centimetres

Argentinosaurus
Argen-tee-no-sore-us • ARGENTINA REPTILE

GROUP	Saurischia, Sauropoda
AGE	90 million years
PERIOD	Upper Cretaceous
LOCALITY	Argentina
LENGTH	up to 30 metres

Avimimus
Aa-vee-meem-us • BIRD MIMIC

GROUP	Saurischia, Theropoda*
AGE	about 75 million years
PERIOD	Upper Cretaceous
LOCALITY	Mongolia
LENGTH	up to 1.5 metres

Bambiraptor
Bambi-rap-tor • BAMBI PLUNDERER

GROUP	Saurischia, Dromaeosauroidea
AGE	76–74 million years
PERIOD	Upper Cretaceous
LOCALITY	USA
LENGTH	about 1 metre

Barosaurus
Bar-oh-sore-us • HEAVY LIZARD

GROUP	Saurischia, Sauropoda
AGE	155–145 million years
PERIOD	Upper Jurassic
LOCALITY	USA
LENGTH	up to 24 metres

Baryonyx
Bar-ee-on-icks • HEAVY CLAW

GROUP	Saurischia, Theropoda*
AGE	about 124 million years
PERIOD	Lower Cretaceous
LOCALITY	England
LENGTH	10.5 metres

Beipaosaurus
Bay-pow-sore-us • BEIPAO LIZARD

GROUP	Saurischia, Maniraptora
AGE	125–120 million years
PERIOD	Lower Cretaceous
LOCALITY	China
LENGTH	up to 3.5 metres

Brachiosaurus
Brak-ee-oh-sore-us • ARM LIZARD

GROUP	Saurischia, Sauropoda
AGE	155–140 million years
PERIOD	Upper Jurassic
LOCALITY	Tanzania, USA
LENGTH	up to 23 metre

Camarasaurus

Brachytrachelopan
Brak-ee-track-el-o-pan • SHORT-NECKED PAN

GROUP	Saurischia, Sauropoda
AGE	147–145 million years
PERIOD	Upper Jurassic
LOCALITY	Argentina
LENGTH	9 metres

Buitreraptor
Beaut-ree-rap-tor • VULTURE ROOST PLUNDERER

GROUP	Saurischia, Dromaeosauroidea
AGE	76–74 million years
PERIOD	Upper Cretaceous
LOCALITY	Argentina
LENGTH	unknown, perhaps up to 2 metres

Camarasaurus
Kam-ahra-sore-us • CHAMBERED LIZARD

GROUP	Saurischia, Sauropoda
AGE	155–145 million years
PERIOD	Upper Jurassic
LOCALITY	USA
LENGTH	up to 20 metres

Camptosaurus
Kamp-toe-sore-us • BENT LIZARD

GROUP	Ornithischia, Ornithopoda
AGE	155–145 million years
PERIOD	Upper Jurassic
LOCALITY	USA
LENGTH	up to 5 metres

Caudipteryx

Carcharodontosaurus
Kar-kar-o-dont-oh-sore-us • CARCHARODON LIZARD

GROUP	Saurischia, Theropoda
AGE	98–94 million years
PERIOD	Upper Cretaceous
LOCALITY	Morocco
LENGTH	15 metres

Carnotaurus
Kar-noh-tore-us • MEAT-EATING BULL

GROUP	Saurischia, Theropoda
AGE	90–?70 million years
PERIOD	Lower/Upper Cretaceous
LOCALITY	Argentina
LENGTH	7.6 metres

Caudipteryx
Cor-dip-ter-ics • TAIL FEATHER

GROUP	Saurischia, Maniraptora
AGE	125–120 million years
PERIOD	Lower Cretaceous
LOCALITY	China
LENGTH	About 1 metre

Centrosaurus
Sent-row-sore-us • HORNED LIZARD

GROUP	Ornithischia, Ceratopia
AGE	76–74 million years
PERIOD	Upper Cretaceous
LOCALITY	Canada
LENGTH	up to 5 metres

Chasmosaurus
Kas-mow-sore-us • CLEFT LIZARD

GROUP	Ornithischia, Ceratopia
AGE	76–74 million years
PERIOD	Upper Cretaceous
LOCALITY	Canada
LENGTH	up to 5 metres

Coelophysis
Seel-oh-fie-sis • HOLLOW FORM

GROUP	Saurischia, Coelurosauria
AGE	225–220 million years
PERIOD	Upper Triassic
LOCALITY	USA
LENGTH	up to 3 metres

Coelurus
Seel-ure-us • HOLLOW TAIL

GROUP	Saurischia, Coelurosauria
AGE	155–145 million years
PERIOD	Upper Jurassic
LOCALITY	USA
LENGTH	up to 2.5 metres

Compsognathus
Komp-sog-nay-thus • PRETTY JAW

GROUP	Saurischia, Coelurosauria
AGE	147 million years
PERIOD	Upper Jurassic
LOCALITY	Germany
LENGTH	65 centimetres

Confuciusornis
Con-few-shush-or-niss • CONFUCIUS BIRD

GROUP	Saurischia, Aviales
AGE	125–120 million years
PERIOD	Lower Cretaceous
LOCALITY	China
LENGTH	up to about 2.5 metres

Corythosaurus
Cor-ith-oh-sore-us • HELMET LIZARD

GROUP	Ornithischia, Ornithopoda
AGE	76–74 million years
PERIOD	Upper Cretaceous
LOCALITY	Canada, USA
LENGTH	up to 10 metres

Craspedodon
Krass-pede-oh-don • EDGE TOOTH

GROUP	Ornithischia, Ornithopoda
AGE	86–83 million years
PERIOD	Upper Cretaceous
LOCALITY	Belgium
LENGTH	unknown

Cryolophosaurus
Cry-oh-loaf-oh-sore-us • FROZEN CRESTED REPTILE

GROUP	Saurischia, Theropoda
AGE	?170 million years
PERIOD	Lower Jurassic
LOCALITY	Antarctica
LENGTH	7–8 metres

Dacentrurus
Dah-sent-rue-rus • POINTED TAIL

GROUP	Ornithischia, Stegosauria
AGE	157–152 million years
PERIOD	Upper Jurassic
LOCALITY	France, England, Portugal
LENGTH	perhaps about 6 metres

Deinocheirus
Dine-oh-kire-us • TERRIBLE HAND

GROUP	Saurischia, Ornithomimosauria
AGE	70–65 million years
PERIOD	Upper Cretaceous
LOCALITY	Mongolia
LENGTH	unknown, arms are 3 metres long

Deinonychus
Die-non-i-kus • TERRIBLE CLAW

GROUP	Saurischia, Dromaeosauroidia
AGE	110 million years
PERIOD	Lower Cretaceous
LOCALITY	USA
LENGTH	up to 3.3 metres

Dilong
Di-long • EMPEROR DRAGON

GROUP	Saurischia, Carnosauria
AGE	about 125 million years
PERIOD	Lower Cretaceous
LOCALITY	China
LENGTH	1.6 metres

Edmontosaurus

Diplodocus
Di-ploh-dok-us • DOUBLE BEAM

GROUP	Saurischia, Sauropoda
AGE	155–145 million years
PERIOD	Upper Cretaceous
LOCALITY	USA
LENGTH	up to 27 metres

Dromaeosaurus
Dro-may-oh-sore-us • RUNNING LIZARD

GROUP	Saurischia, Dromaeosauroidea
AGE	76–74 million years
PERIOD	Upper Cretaceous
LOCALITY	Canada
LENGTH	up to 1.8 metres

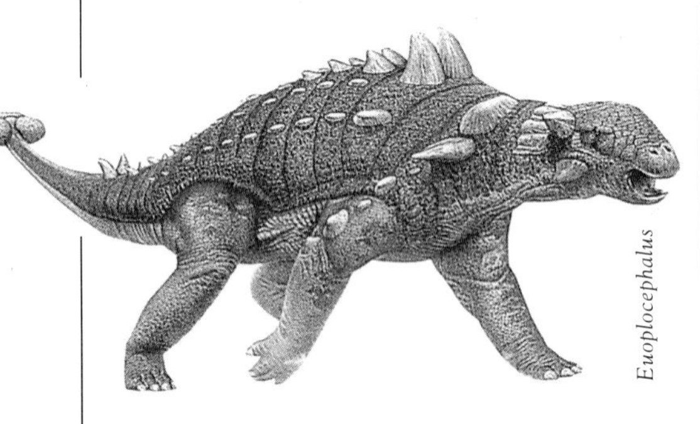

Euoplocephalus

Dryosaurus
Dry-oh-sore-us • OAK LIZARD

GROUP	Ornithischia, Ornithopoda
AGE	155–140 million years
PERIOD	Upper Jurassic
LOCALITY	Tanzania, USA
LENGTH	up to 4 metres

Dryptosaurus
Drip-toe-sore-us • WOUNDING LIZARD

GROUP	Saurischia, Carnosauria
AGE	74–65 million years
PERIOD	Upper Cretaceous
LOCALITY	USA
LENGTH	perhaps about 5 metres

Edmontonia
Ed-mont-own-ee-ah • OF EDMONTON

GROUP	Ornithischia, Ankylosauria
AGE	76–74 million years
PERIOD	Upper Cretaceous
LOCALITY	Canada
LENGTH	about 4 metres

Edmontosaurus
Ed-mont-oh-sore-us • EDMONTON LIZARD

GROUP	Ornithischia, Ornithopoda
AGE	76–65 million years
PERIOD	Upper Cretaceous
LOCALITY	Canada
LENGTH	up to 13 metres

Elasmosaurus (not a dinosaur)
Ee-laz-mow-sore-us • RIBBON LIZARD

GROUP	Plesiosauria
AGE	70–65 million years
PERIOD	Upper Cretaceous
LOCALITY	USA
LENGTH	up to 12 metres

Euoplocephalus
You-op-loh-keff-ah-lus • WELL-ARMOURED HEAD

GROUP	Ornithischia, Ankylosauria
AGE	about 71 million years
PERIOD	Upper Cretaceous
LOCALITY	Canada
LENGTH	up to 6 metres

Euparkeria (not a dinosaur)
You-park-ear-ee-ah • OF PARKER'S

GROUP	Archosauria
AGE	245–241 million years
PERIOD	Lower Triassic
LOCALITY	South Africa
LENGTH	about 1 metre

Eoraptor
Ee-oh-rap-tor • DAWN PLUNDERER

GROUP	Saurischia, Theropoda
AGE	228 million years
PERIOD	Upper Triassic
LOCALITY	Argentina
LENGTH	about 1 metre

Hypsilophodon

Falcarius
Fal-car-ee-us • SICKLE MAKER

GROUP	Saurischia, Maniraptora
AGE	125 million years
PERIOD	Lower Cretaceous
LOCALITY	USA
LENGTH	up to 4 metres

Gallimimus
Gal-lee-meem-us • CHICKEN MIMIC

GROUP	Saurischia, Ornithomimosauria
AGE	74–70 million years
PERIOD	Upper Cretaceous
LOCALITY	Mongolia
LENGTH	up to 5.6 metres

Iguanodon

Giganotosaurus
Gig-an-oh-toe-sore-us • GIANT SOUTHERN LIZARD

GROUP	Saurischia, Theropoda
AGE	90 million years
PERIOD	Upper Cretaceous
LOCALITY	Argentina
LENGTH	up to 12.5 metres

Gilmoreosaurus
Gil-more-oh-sore-us • GILMORE'S LIZARD

GROUP	Ornithischia, Ornithopoda
AGE	80–70 million years
PERIOD	Upper Cretaceous
LOCALITY	China
LENGTH	perhaps about 6 metres

Hadrosaurus
Had-row-sore-us • BIG LIZARD

GROUP	Ornithischia, Ornithopoda
AGE	83–74 million years
PERIOD	Upper Cretaceous
LOCALITY	USA
LENGTH	up to 8 metres

Herrerasaurus
Herr-air-ah-sore-us • HERRERA LIZARD

GROUP	Dinosauria, Theropoda
AGE	230–225 million years
PERIOD	Upper Triassic
LOCALITY	Argentina
LENGTH	up to 3 metres

Hylaeosaurus
High-lay-ee-oh-sore-us • WOODLAND LIZARD

GROUP	Ornithischia, Ankylosauria
AGE	150–135 million years
PERIOD	Lower Cretaceous
LOCALITY	England
LENGTH	up to 4 metres

Hypsilophodon
Hip-sih-loh-foe-don • HIGH RIDGE TOOTH

GROUP	Ornithischia, Ornithopoda
AGE	about 125 million years
PERIOD	Lower Cretaceous
LOCALITY	England
LENGTH	up to 2.3 metres

Iguanodon
Ig-wha-noh-don • IGUANA TOOTH

GROUP	Ornithischia, Ornithopoda
AGE	130–115 million years
PERIOD	Lower Cretaceous
LOCALITY	Belgium, England, Germany, Spain, USA
LENGTH	up to 10 metres

Irritator
Irr-it-ah-tor • IRRITATOR

GROUP	Saurischia, Theropoda
AGE	112–98 million years
PERIOD	Lower Cretaceous
LOCALITY	Argentina
LENGTH	unknown

Janenschia
Yan-ensh-ee-ah • JANENSCH

GROUP	Saurischia, Sauropoda
AGE	155–140 million years
PERIOD	Upper Jurassic
LOCALITY	Tanzania, Malawi
LENGTH	unknown

Kentrosaurus
Kent-row-sore-us • SPIKY LIZARD

GROUP	Ornithischia, Stegosauria
AGE	about 140 million years
PERIOD	Upper Jurassic
LOCALITY	Tanzania
LENGTH	up to 3 metres

Kritosaurus
Krit-oh-sore-us • NOBLE LIZARD

GROUP	Ornithischia, Ornithopoda
AGE	80–75 million years
PERIOD	Upper Cretaceous
LOCALITY	USA
LENGTH	up to 8 metres

Lagosuchus (not a dinosaur)
Lag-oh-sue-kus • RABBIT CROCODILE

GROUP	Dinosauromorpha
AGE	241–235 million years
PERIOD	Middle Triassic
LOCALITY	Argentina
LENGTH	about 30 centimetres

Lambeosaurus
Lam-bee-oh-sore-us • LAMBE'S LIZARD

GROUP	Ornithischia, Ornithopoda
AGE	76–74 million years
PERIOD	Upper Cretaceous
LOCALITY	Canada
LENGTH	up to 9 metres

Leaellynasaura
Lee-ell-in-a-sore-a • LEAELLYN'S LIZARD

GROUP	Ornithischia, Ornithopoda
AGE	115–110 million years
PERIOD	Lower Cretaceous
LOCALITY	Australia
LENGTH	up to 2 metres

Maiasaura
My-ah-sore-ah • GOOD MOTHER LIZARD

GROUP	Ornithischia, Ornithopoda
AGE	80–75 million years
PERIOD	Upper Cretaceous
LOCALITY	USA
LENGTH	up to 9 metres

Majungatholus
Maj-ung-ath-oh-lus • MAJUNGA DOME

GROUP	Saurischia, Theropoda
AGE	?70 million years
PERIOD	Upper Cretaceous
LOCALITY	Madagascar
LENGTH	up to ?9 metres

Ouranosaurus

Mamenchisaurus
Mah-men-chi-sore-us • MAMENCHI LIZARD

GROUP	Saurischia, Sauropoda
AGE	155–145 million years
PERIOD	Upper Jurassic
LOCALITY	China
LENGTH	up to 22 metres

Massospondylus
Mass-oh-spon-di-lus • MASSIVE VERTEBRA

GROUP	Saurischia, Prosauropoda
AGE	208–204 million years
PERIOD	Lower Jurassic
LOCALITY	South Africa
LENGTH	up to 4 metres

Megalosaurus
Meg-ah-low-sore-us • BIG LIZARD

GROUP	Saurischia, Carnosaura
AGE	170–155 million years
PERIOD	Jurassic
LOCALITY	England, Tanzania
LENGTH	up to 9 metres

Mei
May-i • SOUNDLY SLEEPING

GROUP	Saurischia, Maniraptora
AGE	about 125 million years
PERIOD	Lower Cretaceous
LOCALITY	China
LENGTH	about 30 centimetres

Microraptor
My-crow-rap-tor • LITTLE PLUNDERER

GROUP	Saurischia, Dromaeosauroidea
AGE	about 125 million years
PERIOD	Lower Cretaceous
LOCALITY	China
LENGTH	77 centimetres (the tail was more than half the total length)

Ornitholestes
Orn-ith-oh-lest-eez • BIRD ROBBER

GROUP	Saurischia, Coelurosauria
AGE	155–145 million years
PERIOD	Upper Jurassic
LOCALITY	USA
LENGTH	up to 2 metres

Ornithopsis
Orn-ith-op-sis • BIRD-LIKE STRUCTURE

GROUP	Saurischia, Sauropoda
AGE	about 125 million years
PERIOD	Lower Cretaceous
LOCALITY	England
LENGTH	unknown, perhaps 20 metres

Orodromeus
Ore-oh-drom-ee-us • MOUNTAIN RUNNER

GROUP	Ornithischia, Ornithopoda
AGE	about 74 million years
PERIOD	Upper Cretaceous
LOCALITY	USA
LENGTH	up to 2 metres

Ouranosaurus
Oo-ran-oh-sore-us • BRAVE MONITOR LIZARD

GROUP	Ornithischia, Ornithopoda
AGE	about 115 million years
PERIOD	Lower Cretaceous
LOCALITY	Niger
LENGTH	up to 7 metres

Oviraptor
Oh-vee-rap-tor • EGG THIEF

GROUP	Saurischia, Oviraptorosauria
AGE	85–75 million years
PERIOD	Upper Cretaceous
LOCALITY	Mongolia
LENGTH	up to 1.8 metres

Parasaurolophus

Pachycephalosaurus

Pack-ee-keff-ah-low-sore-us • THICK-HEADED LIZARD

GROUP	Ornithischia, Pachycephalosauria
AGE	about 67 million years
PERIOD	Upper Cretaceous
LOCALITY	USA
LENGTH	up to 8 metres

Pachyrhinosaurus

Pack-ee-rine-oh-sore-us • THICK-NOSED LIZARD

GROUP	Ornithischia, Ceratopia
AGE	76–74 million years
PERIOD	Upper Cretaceous
LOCALITY	Canada, USA
LENGTH	up to 6 metres

Parasaurolophus

Par-ah-sore-ol-oh-fus • LIKE SAUROLOPHUS

GROUP	Ornithischia, Ornithopoda
AGE	76–74 million years
PERIOD	Upper Cretaceous
LOCALITY	Canada, USA
LENGTH	up to 10 metres

Parksosaurus

Parks-oh-sore-us • PARKS' LIZARD

GROUP	Ornithischia, Ornithopoda
AGE	76–74 million years
PERIOD	Upper Cretaceous
LOCALITY	Canada
LENGTH	up to 3 metres

Pelorosaurus

Pel-oh-row-sore-us • MONSTROUS LIZARD

GROUP	Saurischia, Sauropoda
AGE	about 150 million years
PERIOD	Upper Jurassic
LOCALITY	England
LENGTH	unknown, perhaps 20–25 metres

Plateosaurus

Plat-ee-oh-sore-us • FLAT LIZARD

GROUP	Saurischia, Prosauropoda
AGE	about 210 million years
PERIOD	Upper Triassic
LOCALITY	France, Germany, Switzerland
LENGTH	up to 7 metres

Psittacosaurus

Scelidosaurus

Polacanthus

Pol-a-kan-thus • MANY SPINED

GROUP	Ornithischia, Ankylosauria
AGE	about 125 million years
PERIOD	Lower Cretaceous
LOCALITY	England
LENGTH	up to 5 metres

Protarchaeopteryx

Pro-toe-ark-ee-op-ter-ics • BEFORE ARCHAEOPTERYX

GROUP	Saurischia, Maniraptora
AGE	about 125 million years
PERIOD	Lower Cretaceous
LOCALITY	China
LENGTH	about 1 metre

Protoceratops

Pro-toe-ker-ah-tops • FIRST HORNED FACE

GROUP	Ornithischia, Ceratopia
AGE	85–80 million years
PERIOD	Upper Cretaceous
LOCALITY	Mongolia
LENGTH	up to 1.8 metres

Psittacosaurus

Sit-ak-oh-sore-us • PARROT LIZARD

GROUP	Ornithischia, Ceratopia
AGE	124–97 million years
PERIOD	Upper Cretaceous
LOCALITY	China, Mongolia, Russia
LENGTH	up to 1.8 metres

Quetzalcoatlus (not a dinosaur)

Kwet-zal-co-art-lus • FOR QUETZALCOATL

GROUP	Pterosauria
AGE	67–65 million years
PERIOD	Upper Cretaceous
LOCALITY	USA
LENGTH	wing span up to 11 metres

Rhabdodon

Rab-doe-don • ROD TOOTH

GROUP	Ornithischia, Ornithopoda
AGE	83–70 million years
PERIOD	Upper Cretaceous
LOCALITY	Austria, France, Spain, Romania
LENGTH	up to 3 metres

Saurolophus

Sore-ol-oh-fus • RIDGED LIZARD

GROUP	Ornithischia, Ornithopoda
AGE	74–70 million years
PERIOD	Upper Cretaceous
LOCALITY	Canada, Mongolia
LENGTH	up to 12 metres

Saurornithoides

Sore-orn-ith-oy-deez • BIRD-LIKE LIZARD

GROUP	Saurischia, Dromaeosauroidea
AGE	80–74 million years
PERIOD	Upper Cretaceous
LOCALITY	Canada
LENGTH	up to 2 metres

Scartopus

Skar-toe-pus • NIMBLE FOOT

GROUP	Saurischia, Coelurosauria
AGE	about 95 million years
PERIOD	Upper Cretaceous
LOCALITY	Australia
LENGTH	unknown

Scelidosaurus

Skel-ide-oh-sore-us • LIMB LIZARD

GROUP	Ornithischia, Thyreophora
AGE	203–194 million years
PERIOD	Lower Jurassic
LOCALITY	England
LENGTH	up to 4 metres

Segnosaurus

Seg-no-sore-us • SLOW LIZARD

GROUP	Saurischia, Maniraptora
AGE	97–88 million years
PERIOD	Upper Cretaceous
LOCALITY	Mongolia
LENGTH	up to 4 metres

Seismosaurus

Size-moh-sore-us • EARTH-SHAKING LIZARD

GROUP	Saurischia, Sauopoda
AGE	155–145 million years
PERIOD	Upper Jurassic
LOCALITY	USA
LENGTH	?40 metres

Sinornithosaurus

Sine-or-nith-oh-sore-us • CHINESE BIRD-LIKE DINOSAUR

GROUP	Saurischia, Theropoda
AGE	122–120 million years
PERIOD	Lower Cretaceous
LOCALITY	China
LENGTH	more than 1 metre

Sinosauropteryx

Sine-oh-sore-op-ter-iks • CHINESE DRAGON WING

GROUP	Saurischia, Theropoda
AGE	122–120 million years
PERIOD	Lower Cretaceous
LOCALITY	China
LENGTH	0.68–1.25 metres

Spinosaurus

Spine-oh-sore-us • THORN LIZARD

GROUP	Saurischia, Theropoda
AGE	95–90 million years
PERIOD	Upper Cretaceous
LOCALITY	Northern Africa
LENGTH	15–?20 metres

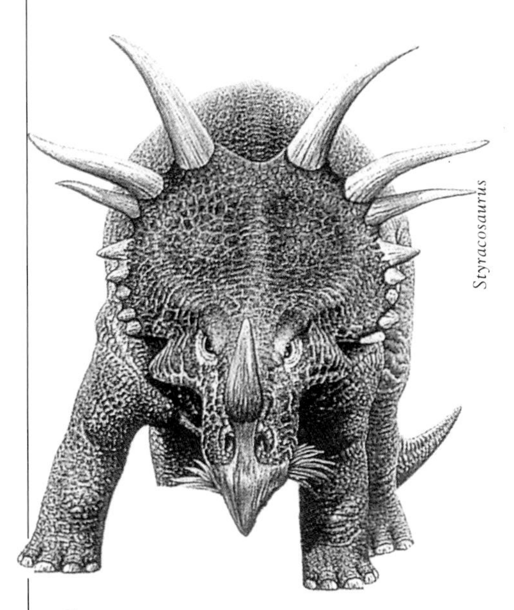

Styracosaurus

Stegosaurus

Steg-oh-sore-us • ROOF LIZARD

GROUP	Ornithischia, Stegosauria
AGE	155–145 million years
PERIOD	Upper Jurassic
LOCALITY	USA
LENGTH	up to 9 metres

Struthiosaurus

Strew-thee-oh-sore-us • OSTRICH LIZARD

GROUP	Ornithischia, Ankylosauria
AGE	83–75 million years
PERIOD	Upper Cretaceous
LOCALITY	Austria, Romania
LENGTH	up to 2 metres

Styracosaurus

Sty-rack-oh-sore-us • SPIKED LIZARD

GROUP	Ornithischia, Ceratopia
AGE	85–80 million years
PERIOD	Upper Cretaceous
LOCALITY	Canada, USA
LENGTH	up to 5.5 metres

Suchomimus

Sook-oh-me-mus • CROCODILE MIMIC

GROUP	Saurischia, Theropoda
AGE	about 100 million years
PERIOD	Lower Cretaceous
LOCALITY	Niger
LENGTH	up to 11 metres

Tenontosaurus

Ten-on-toe-sore-us • SINEW LIZARD

GROUP	Ornithischia, Ornithopoda
AGE	110 million years
PERIOD	Lower Cretaceous
LOCALITY	USA
LENGTH	up to 6.5 metres

Therizinosaurus

There-iz-ine-oh-sore-us • SCYTHE LIZARD

GROUP	Saurischia, Maniraptora
AGE	74–70 million years
PERIOD	Upper Cretaceous
LOCALITY	Mongolia
LENGTH	up to 12 metres

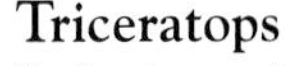

Triceratops

Try-ker-ah-tops • THREE-HORNED FACE

GROUP	Ornithischia, Ceratopia
AGE	67–65 million years
PERIOD	Upper Cretaceous
LOCALITY	USA
LENGTH	up to 9 metres

Troodon

True-oh-don • WOUNDING TOOTH

GROUP	Saurischia, Dromaeosauroidea
AGE	75–70 million years
PERIOD	Upper Cretaceous
LOCALITY	Canada, USA
LENGTH	up to 2.4 metres

Tuojiangosaurus

Two-yang-oh-sore-us • TUOJIANG LIZARD

GROUP	Ornithischia, Stegosauria
AGE	157–154 million years
PERIOD	Upper Jurassic
LOCALITY	China
LENGTH	up to 6.5 metres

Tyrannosaurus

Tie-ran-oh-sore-us • TYRANT LIZARD

GROUP	Saurischia, Carnosauria
AGE	67–65 million years
PERIOD	Upper Cretaceous
LOCALITY	USA
LENGTH	up to 12 metres

Utahraptor

Yew-taa-rap-tor • UTAH ROBBER

GROUP	Saurischia, Dromaeosauroidea
AGE	about 125 million years
PERIOD	Lower Cretaceous
LOCALITY	USA
LENGTH	up to 7 metres

Velociraptor

Vel-oss-ee-rap-tor • QUICK PLUNDERER

GROUP	Saurischia, Dromaeosauroidea
AGE	84–80 million years
PERIOD	Upper Cretaceous
LOCALITY	China, Mongolia
LENGTH	up to 1.8 metres

Tyrannosaurus

GLOSSARY

anatomy — the structure of an animal.
angiosperms — flowering plants.
ankylosaurs — armoured ornithischian dinosaurs.
archosaurs — a major group of reptiles which includes the living crocodiles, and dinosaurs, pterosaurs and thecodontians which are all extinct.
bipedal — walking on two legs.
carnivore — meat-eater.
carnosaurs — large, heavily-built theropod dinosaurs.
ceratopians — four-legged ornithischian dinosaurs with horns and neck frills.
coelurosaurs — small to medium-sized slenderly built theropod dinosaurs.
conifers — cone-bearing trees such as pines and firs.
Cretaceous — the third and last Period in the Mesozoic Era, from 145 to 65 million years ago.
cycads — short-trunked plants with rather palm-like leaves which were common in the Mesozoic Era.
dinosaurs — land-dwelling reptiles with upright stance that lived between 230 and 65 million years ago.
evolution — gradual change in the characteristics of animals and plants brought about by natural selection acting on successive generations.
extinction — dying out of a species.
fossil — remains of once living organisms that are preserved usually by burial and chemical change.
gastroliths — stomach stones used for grinding up food.
Gondwana — southern hemisphere supercontinent that existed in Triassic times incorporating Africa, Antarctica, Australia, India, Madagascar and South America.
hadrosaurs — duck-billed ornithopod dinosaurs with multiple rows of grinding teeth.
herbivore — plant-eater.
ichthyosaurs — marine reptiles with a streamlined, superficially dolphin-like appearance that lived in the seas during the Mesozoic Era.
Jurassic — the middle Period in the Mesozoic Era, from 208 to 145 million years ago.
Laurasia — northern hemisphere supercontinent which existed in Triassic times incorporating Asia, Europe and North America.
ligaments — bundles of tough fibrous tissue which link together the bones of a skeleton and support the joints.
mammals — backboned animals that have hair and feed their young on milk.
mammal-like reptiles — group of reptiles that lived mainly in Triassic and earlier times from which mammals evolved.
Mesozoic — 'Middle life', the major time span (era) 245–65 million years ago, informally known as 'the age of dinosaurs'.
ornithischians — dinosaurs with a hip structure in which the two lower bones on each side lie parallel and point backwards; all ornithischians were plant-eaters.
ornithopods — primarily bipedal ornithischian dinosaurs that developed specialized teeth to grind up tough vegetation.
pachycephalosaurs — bipedal ornithischian dinosaurs with thickened 'dome-headed' skulls.
palaeontologist — a person who studies fossils.
Pangea — the single land mass formed by the collision of all the continents. This happened long before the earliest dinosaurs appeared by which time Pangea had begun to separate into Laurasia and Gondwana.
plesiosaurs — fish-eating marine reptiles with four flippers that lived in the Mesozoic Era.
prosauropods — one of the earliest groups of dinosaurs, these long-necked saurischian plant-eaters with large thumb claws were distributed world-wide and died out in the early Jurassic.
pterosaurs — flying reptiles that lived during the Mesozoic Era; they were close relatives of the dinosaurs.
quadrupedal — walking on four legs.
reptile — backboned animal that is 'cold-blooded', has a scaly skin and lays eggs.
saurischians — dinosaurs with a hip structure in which the two lower bones on each side point in opposite directions.
sauropods — large quadrupedal saurischian plant-eating dinosaurs, with very long necks and tails.
scavenger — meat-eater that feeds on dead animals, sometimes by stealing kills from a hunter.
spine — backbone (see under vertebra).
stegosaurs — plant-eating dinosaurs with tall plates running along their backs and two pairs of tail spikes.
tendon — tough inelastic tissue that joins muscles to bones.
thecodontians — early archosaurs from which crocodiles, dinosaurs and pterosaurs evolved during the Triassic Period.
theropods — a group of saurischian dinosaurs that includes all the meat-eaters, almost all were bipedal.
trackway — a series of footprints left as an animal walks over soft ground.
Triassic — the first Period in the Mesozoic Era, from 245 to 208 million years ago.
vertebra — a single bone from the back. A chain of vertebrae make up the backbone.

FURTHER READING

General Books about Dinosaurs

Barrett, P. M.
National Geographic Dinosaurs
Washington D. C., National Geographic, 2001

Currie, P. J. and Padian, K.
Encyclopedia of Dinosaurs
San Diego, Academic Press, 1997

Farlow, J. O. and Brett-Surman, M. K.
The Complete Dinosaur
Bloomington, Indiana University Press, 1997

Fastovsky, D, E. and Weishampel, D. B.
The Evolution and Extinction of the Dinosaurs. 2nd Edition
New York, Cambridge University Press, 2005

Milner, A.
Dino-Birds. From Dinosaurs to Birds
London, The Natural History Museum, 2002

Norman, D. B. and Wellnhofer, P.
The Illustrated Encyclopedia of Dinosaurs and Pterosaurs: an Original and Compelling Insight into Life in the Dinosaur Kingdom.
London, Salamander Books, 2000
(combined reprint of Norman, D. B. *The Illustrated Encyclopedia of Dinosaurs*, 1985 and Wellnhofer, P. *The Illustrated Encyclopedia of Pterosaurs*, 1991.)

Books on Special Topics

Archibald, J. D.
Dinosaur Extinction and the End of an Era. What the Fossils Say
New York, Columbia University Press, 1996

Benton, M. J.
The Reign of the Reptiles
London, Kingfisher Books, 1990

Benton, M. J.
Vertebrate Palaeontology: Biology and Evolution, 3rd edition
London, Chapman and Hall, 2004

Carpenter, K., Hirsch, K. F. and Horner, J. R.
Dinosaur Eggs and Babies
New York, Cambridge University Press, 1994

Croucher, R. and Woolley, A. R.
Fossils, Minerals and Rocks. Collection and Preservation
London, British Museum (Natural History)/Cambridge University Press, 1982

Currie, P. J., Koppelhaus, E. B., Shugar, M. A. and Wright, J. L. (Eds)
Feathered Dragons, Studies on the Transition from Dinosaurs to Birds.
Bloomington, Indiana University Press, 2004

Lockley, M. G.
Tracking Dinosaurs. A New Look at an Ancient World
Cambridge, Cambridge University Press, 1991

Lockley M. and Meyer, C.
Dinosaur Tracks and Other Fossil Footprints of Europe
New York. Columbia University Press, 2000

INDEX

Note: **Bold** refers to captions to illustrations

Picture Acknowledgements cover, p.3 NHM/Kokoro Dreams; pp.8-9 John Sibbick/NHM; p.10 Jenny Halstead-Imagination; p.12 Andrew Milner; pp.13-15 Ray Burrows/NHM; p.14 (tr) John Sibbick/NHM; p.16 (l) Sally Dray, (r) © 1991, Studio Editions reprinted from *The Dinosaur Wall Chart* by permission of the publisher; pp.20-1 John Sibbick/NHM; p.23 (l) Imagination; (r) Angela Milner; p. 24 (tl), (mtl) Andrew Milner, (mbl) Gunther Ziesler, (bl) Wayne Lankinen/Bruce Coleman; p.25 Alligator Books Ltd; p.26 (b) Imagination; p.27 (tl) E. & P. Bauer/Wildphotos; p.28 (tl) Jenny Halstead, (b) John Sibbick/NHM; p.29 © The Field Museum, GN89714_2RDC; p.30 (t) Jim Farlow, artist Jim Whitcraft; (r) J. & D. Bartlett/Bruce Coleman; p.31 (r) Imagination; pp.32-3 John Sibbick/NHM; p.35 (b) Kent Stevens; p.37 (t) Jenny Halstead; p.39 (tm) Peter Johnson/Corbis, (bm) William S. Paton; p.41 (b) © Dr Emily Rayfield; p.46 (t) Gerald Cubitt, (b) Pictor; p.47 (t) John Sibbick/NHM; p.47 (b) Ray Burrows/NHM; p.48 (l) Alligator Books Ltd; p.49 John Sibbick/NHM; pp.50-1 Peabody Museum, Yale University; p.52 (l) Jenny Halstead, (tr) Austin James Stevens/Bruce Coleman, (mr) Bruce Davidson/Naturepl.com, (br) Rod Williams/Naturepl.com; p.53 Alligator Books Ltd; p.54-7 (t) John Sibbick/NHM; p.57 (b) Rafi Ben-Shahar/OSF; p.58 (t) John Sibbick/NHM; (m) Royal Tyrell Museum/Alberta Culture and Multiculturalism; p.59-p.60 (tl) John Sibbick/NHM; p.60 (tr) Jenny Halstead, (b) Louie Psihoyos/Corbis; p.61 (t) Mark Norell, (b) Mick Ellison; p.65-7 (t) John Sibbick/NHM; p.69 (r) Ray Burrows/NHM; p.70 (l) Jenny Halstead; p.71 John Sibbick/NHM; p.73 (b), p.75 (b) Jenny Halstead; (t) Christopher Brochu; p.76 R. E. H. Reid; p.81 (bl) Eric Albert, (br) Michael Long/NHM; p.82 © Douglas Henderson from *Dinosaurs - A Global View* by Sylvia and Stephen Czerkau, Dragon's World Ltd; p.83 (tl) Glen A. Izett/US Geological Survey, (tr) Harry Foster reproduced with the permission of the Canadian Museum of Nature, Ottawa, (b) © Royal Observatory, Edinburgh; p.85 (t) US Geological Survey, (b) Ray Burrows/NHM; p.86 (l) Hjalmar R. Bardarson/OSF, (r) K. G. Cox, Oxford University; p. 87 © Douglas Henderson from *Dinosaurs - A Global View* by Sylvia and Stephen Czerkau, Dragon's World Ltd; p.91 (l) The Dover Pictorial Archive Series, (r) From *Dinosaurs from China*, China Ocean Press; pp.96-7 Dept. of Palaeontology, Royal Belgian Institute of Natural Sciences, Brussels; p.100 (main) Museum für Naturkunde, Tölke, Berlin; p.101 (tr) Neg. 410764, Courtesy Dept, Library Services, American Museum of Natural History, (ml and b) Mee-Mann Chang/Institute of Vertebrate Palaeontology and Palaeoanthropology, Beijing; p.105 Lisa Wilson/NHM, adapted with permission from Macmillan Publishers Ltd, *Nature*, 435, Rauhut *et al*, Discovery of a short-necked sauropod dinosaur from the Late Jurassic Period of Patagonia © 2005; p.106 Firecrest Books Ltd; pp.107 © 1992, Hanna-Barbera Productions, Inc. Licensed by C. P. L.; p.110 Ray Burrows/NHM; p.112 (tl) Andrew Milner; p.113 (b) John Holmes; p.115 John Sibbick/NHM; p.117 Reprinted with permission from *Nature* vol. 324, 27 Nov. 1986, Macmillan Publishers Ltd; pp.118-19 Ray Burrows/NHM; p.121 (b) John Holmes; p.124-6 John Sibbick/NHMPL; p.127, p.129 (t) Geological Museum of China/NHMPL; p.129 (b) John Sibbick/NHMPL; p.130 Geological Museum of China/NHMPL; p.131 John Sibbick/NHMPL; p.132 Reproduced by permission from Macmillan Publishers Ltd, *Nature*, 421, X. Zu *et al*, Four-winged Dinosaurs from China © 2005; p.132 (t) John Sibbick/NHMPL; p.133 Geological Museum of China/NHMPL; p.135 (l) John Sibbick/NHMPL; pp.134-140 Alligator Books Ltd. All other pictures are from the Natural History Museum, London. (Key: t=top; m=middle; l=left; r=right; b=bottom; NHM=Natural History Museum). Every effort has been made to contact and accurately credit all copyright holders. If we have been unsuccessful we apologise and welcome corrections for future editions.